BY THERESA MACPHAIL

Allergic

The Viral Network

Allergic

Allergic

OUR IRRITATED BODIES
IN A CHANGING WORLD

● ● ●

Theresa MacPhail

RANDOM HOUSE

NEW YORK

Published in the United States by Random House, an imprint and
division of Penguin Random House LLC, New York.

RANDOM HOUSE and the HOUSE colophon are registered trademarks
of Penguin Random House LLC.

Library of Congress Cataloging-in-Publication Data
Names: MacPhail, Theresa, author.
Title: Allergic: our irritated bodies in a changing world /
Theresa MacPhail.
Description: First edition. | New York: Random House, [2023] |
Includes index.
Identifiers: LCCN 2022032275 (print) | LCCN 2022032276 (ebook) |
ISBN 9780593229194 (hardcover) | ISBN 9780593229200 (ebook)
Subjects: LCSH: Allergy—History. | Allergy—Popular woks.
Classification: LCC RC584 .M28 2023 (print) | LCC RC584 (ebook) |
DDC 616.97/3—dc23/eng/20220916
LC record available at https://lccn.loc.gov/2022032275
LC ebook record available at https://lccn.loc.gov/2022032276

Printed in the United States of America on acid-free paper

randomhousebooks.com

2 4 6 8 9 7 5 3 1

FIRST EDITION

Book design by Simon M. Sullivan

CONTENTS

———————

PART THREE

Treatments

Everything That Irritates Us

———

On August 25, 1996, my dad was cruising down Main Street in our small New Hampshire town in the respectable, boxy four-door sedan that he used to make sales calls during the week. He and his longtime girlfriend, Patricia, were headed out to the beach to enjoy the day surfside. It was 11:20 A.M. and as the sun marched toward its apogee, the temperature slowly rose with it. The car's windows were rolled down, which was typical for my dad. He was an enthusiastic smoker of Marlboro Lights who also eschewed the use of air-conditioning unless it was blisteringly hot. We were New Englanders, after all, and expected to tough out all but the most miserable of weather.

My dad's hand dangled outside his car, a lit cigarette pinched between his fingers, his forearm resting on the warm metal of the car door. The radio was tuned in to an AM station covering the Boston Red Sox. My dad could never get enough baseball. He dialed in to seemingly every game and if one wasn't being played, then he liked to listen to analysis of past matchups and predictions about future ones. As a teenager who was more into reading Dickens and obsessing over Duran Duran, I found his enthusiasm for sports exasperating, especially his addiction to sports radio. As a rule, I would sit in the back seat, trying to concentrate on reading, my eye rolls partially hidden behind a thick paperback book. Sometimes, just to annoy him, I would root for the opposing team until he threatened to pull the car over and let his only child walk home.

But in 1996, I was twenty-four years old. That Sunday in August, I wasn't in the car with my father. I heard about what happened from three sources: the state police, who informed me as next of kin that he was dead; a local funeral director I phoned to see where my father had been taken, who remembered his colleagues discussing the unusual condition of his body; and, twenty-five years later, Patricia, during our first conversation since my father's wake. Yet my dad was such a creature of habit that I have no trouble picturing the events as they likely unfolded. If I close my eyes, I can see him sitting in his car, a Styrofoam cup of hot coffee wedged into the cup holder, his hand loosely resting on top of the wheel.

Growing up, I had a strained relationship with my father. My parents divorced when I was just two months old, and I saw him only a few times throughout my young childhood. The tension that existed between us deepened after my mother's death in a car crash in 1986, when, at fourteen, I moved from my hometown in rural Indiana to live with him and Patricia in suburban New Hampshire. My dad and I were what I euphemistically liked to call "estranged" whenever I tried to explain our familial situation to new acquaintances or friends. I had a father and I loved him; I just never spoke to him.

As my dad drove that day, a solitary bee was on its usual pollen-collecting rounds when its flight trajectory intersected with my father's open car window. The bee became confused and panicked. It stung my dad in the side of his neck, close to his ear. My dad, surprised but still calm, continued to drive.

What happened next was not visible to the naked eye. Events shifted to the microscopic level, inside my father's body. Biology took over.

The bee's stinger introduced its venom—a mixture of water, histamine, pheromones, enzymes, and various amino acids, or proteins—underneath the thin layer of skin into the fatty tissue of my father's neck. Packed tight with blood vessels, the neck is a great circulatory site, so the venom had a unique opportunity to spread rapidly throughout my dad's body. Some of my father's immune cells—his mast cells and basophils—swiftly detected certain of the venom's components.

White blood cells like mast cells and basophils are produced in our bone marrow and circulate throughout the human body, helping to fight off infection or disease by ingesting foreign or harmful materials like viruses, bacteria, and cancer cells. Mast cells can be found in the connective tissues under our skin, lining our respiratory tract and intestines, and in the tissue around our lymph nodes, nerves, and blood vessels. Basophils are found in our bloodstream. Mast cells and basophils, then, are nearly everywhere inside the human body. Their job, to dramatically simplify it, is both to begin and to amplify the severity of our immune response. Think of them as the conductors of our immune system, modulating its response by releasing various proteins and chemicals.

Bee venom is not a natural substance that the human body responds particularly well to, even under normal circumstances in a nonallergic person. Bee venom is naturally hemorrhagic, meaning that it has the ability to blow apart our blood cells. Even so, bee and wasp venoms are relatively harmless in most humans, apart from causing a painful, localized swelling near the injection sites. Everyone's immune cells react to venom; my father's dramatically overreacted, sending his immune system into the deadly spiral known as anaphylaxis. Anaphylaxis is medically defined by the World Health Organization as "a severe, life-threatening systemic hypersensitivity reaction characterized by being rapid in onset with potentially life-threatening airway, breathing, or circulatory problems." What that means in layman's terms is that my father had an underlying allergy to the bee's venom, a hypersensitivity that he tragically underestimated until it was too late.

Just a few weeks prior, while in the parking lot of Walmart, my father had been stung by another bee. When he returned home, he told Patricia he wasn't feeling well and took some Benadryl—a well-known brand of antihistamine, commonly recommended for coping with milder allergic responses. Soon after, he felt better, but Patricia nagged him to see a doctor, suspecting he had an allergy to bees. My father, notoriously bad at taking care of his own physical health (he smoked too much, drank too much bourbon, and ate too many servings of prime rib), demurred.

Allergic responses can strengthen over time with repeated exposures. The first time my father was stung, he may not have had more than a small welt at the site of the sting itself. The second or third time, his body's immune cells would have remembered the offending substances and reacted more swiftly and strongly—causing a proportionally larger reaction. My father's body, unbeknownst to him, was already primed to betray him.

The process of anaphylaxis starts as soon as an antigen—a fancy term for any substance, like bee venom, that initiates an immune response—encounters and activates the mast cells and basophils in your body. My dad's mast cells and basophils began the process of anaphylaxis mere seconds after he was stung in his sedan, as soon as they came into direct contact with the venom's proteins and started emitting histamine. Histamine, an organic compound created by the body, is a key part of a normal immune response, released when cells are injured or stressed. It causes blood vessels to dilate and their walls to become more permeable—thereby making it much easier for infection-fighting white blood cells to leak out of blood vessels and move into affected areas. Histamine is also a signal to other nearby cells to release even more histamine. Think of histamine as the body's chemical alarm system; once it goes off, it alerts your entire immune system to go into action. How does this alarm system feel within your body? Histamine acts on receptors on your organs, causing inflammation, flushing, itching, hives, and swelling.

Unfortunately for my father, everything that happened next would be accelerated simply because he was still sitting upright in his car, his body's position partially obstructing the flow of deoxygenated blood returning back to his heart. The allergic surge of histamine circulating throughout his body caused my dad's veins to dilate too quickly, reducing his blood pressure and the flow of blood to his heart even further, a process that can—and, in my father's case, eventually did—

culminate in cardiac arrest. The excess of histamine additionally shifted fluid from his vascular system—the network of blood vessels throughout his body—into his tissues, causing my father's body, including his neck, to swell. In an effort to help protect the lower airways from inhaled irritants, histamine also thickens mucus, increases mucus production, and causes the smooth muscle tissue around your lungs to tighten. During an anaphylactic event, your airways begin to constrict within minutes. My father, sensing all of this beginning to happen to him, pulled over to the side of the road and asked Patricia to drive.

Panicked and miles away from the nearest hospital, Patricia decided to drive to a local pharmacy for more immediate help. Now in the passenger's seat, my father started to gasp for air, his face changing color.

Minutes later, Patricia pulled the sedan into the tiny lot in front of the small drugstore, threw it into park, and ran for help. The pharmacist on duty that day explained that he couldn't give my father a potentially lifesaving shot of epinephrine, also known as adrenaline, because my father didn't have a current prescription for it. Epinephrine, a natural hormone secreted by the adrenal glands during times of stress, helps to stop the process of anaphylaxis by halting the release of histamine and constricting the blood vessels—thereby aiding blood flow. It also binds to receptors on the smooth muscles of the lungs, helping them to relax and allowing the breath to return to normal. An emergency shot delivers a much greater dose of adrenaline to the body than it can produce on its own in a short amount of time. But instead of administering the drug to my father, the pharmacist called for the paramedics.

When the ambulance finally arrived, emergency medical technicians (EMTs) intubated my father, who could no longer breathe due to the swelling in his neck tissue coupled with the constriction of his lungs. The ambulance did not have any adrenaline on hand, and the pharmacist continued to adamantly, if regretfully, refuse to give the EMTs access to the drug my father now so desperately needed. Despite how cruel his decision may appear to us now, the pharmacist's hands were legally tied. In the 1990s, pharmacists were not allowed to administer adrenaline, even in the case of an emergency. As precious minutes ticked by, my father's body went into shock, the final stage of what is referred to as an inflammatory cascade.

As my father was being loaded into the back of the ambulance, Patricia, hovering over him, asked him to blink if he could still hear her. He softly closed and opened his eyes. She squeezed his hand, still terrified, yet relieved and hopeful. As she climbed back into my father's sedan to drive herself to the emergency room, she listened to the sound of his ambulance's siren as it faded into the distance.

On the way to the hospital, despite all efforts to save him, my father's heart stopped.

James MacPhail—die-hard Boston sports fan, computer chip salesman, Vietnam veteran, Jackie Gleason look-alike, the life of every party, loving son, stand-up comedy aficionado, musical lover, and my father—was gone.

———

As I was researching this book, I turned forty-seven, the same age my father was when he died, and I often found myself thinking about his unusual death as I talked to experts across the country about the puzzle of allergies. Deadly anaphylactic reactions to bee stings remain incredibly rare. Each year, around 3 percent of adults will experience a life-threatening reaction to an insect sting (bees, wasps, or hornets), yet most will survive.[1] In the two decades since my father's death, an average of just sixty-two Americans, or 0.00000002 percent of the general population, have died annually from an insect sting.[2] My dad's death was an outlier, an unfortunate accident, and a life-changing event for all of his friends and family.

But the more I learned about allergy, the more I kept wondering, *Why him?* Was it something about his genetic makeup (and thus also part of my own) that primed his immune system to overreact in the first place? Or was it something about his environment growing up in Boston or the way he had lived his life? In theory, my dad could have become more sensitized to bee venom after being repeatedly stung—

either in childhood or during his two tours of duty in Vietnam. Or, he could have just been very, very unlucky to die from his second encounter with venom in just under a month. Yet as I write this—having finished my research and now three years older than him—I know that there is no way to know for certain what caused my father's allergy because allergies themselves are complicated.

From a biological perspective, I can explain exactly what happened during my father's last moments on earth. The underlying biology is, in many ways, the easiest part of the story to understand and to tell: My father's immune system response was too effective for his own good. In Greek, anaphylaxis literally means "backward defense." My dad's immune system—built to protect him—was completely functional but overly sensitive, misrecognizing a naturally occurring, relatively innocuous substance as a direct threat. Once a heightened immune system reaction begins, it can be nearly impossible to stop. For the people who live with a severe allergy, the paradox of having such a strong, active immune system is that, in addition to protecting you from germs and parasites, it can kill you. And that's exactly what happened to my dad.

The thing that I continue to struggle with—that I simply cannot grasp—is what my dad must have been thinking and feeling as he helplessly watched his own body fail him. How frightened he must have been in those first few seconds as he felt his throat begin to swell shut and his lung muscles contract, cutting off his ability to breathe. How terrified he must have been as his heart began to slow inside his rib cage. What is it like to incrementally, and yet swiftly, die as your immune system goes into overdrive? Would he even have understood what was happening to him? At the very end, as his heart stopped, did he have time to think once more of me, or my grandmother, or his girlfriend? Did he know how much we would miss him?

———

Strange as it may seem, I didn't originally set out to research the topic of allergies because of my father. Over time, I had normalized his death and ruminated on it less and less. For years, the only time I thought about my dad's final moments was when I was sitting outside at a picnic table or walking through a garden and heard a familiar buzzing sound. The mere sight of a bee could send my heart pounding and freeze me in my tracks. But outside of these random encounters with wasps, hornets, or bees, I didn't think all that much about allergies. Until, that is, I was diagnosed with them myself.

In 2015, I was a busy new assistant professor teaching a full course load and trying to write a book about influenza. Ironically, I kept getting sick. Very sick. After being diagnosed with my fourth respiratory infection in less than a year, my doctor shipped me off to see an otolaryngologist—an ear, nose, and throat specialist—proclaiming that something must be wrong with my nasal "plumbing." The otolaryngologist listened to my complaints, examined my doctor's notes, and then looked inside my nasal cavities and down my throat with a scope.

"You've got some serious irritation," he said, still peering into the deep recesses of my nose. "Much more than would occur with just an infection. I'd say that you have allergies. That's your real problem."

This was complete news to me. I had never suffered from undue sneezing or sniffling; I had no red or puffy eyes, no itching or redness or tingling of my skin, no upset stomachs. As far as I knew, I was allergy-free. Except that here was a specialist, someone with years of clinical experience, telling me that, actually, I was one of the millions of allergy sufferers living in the United States. And those allergies were making it more difficult for my overwhelmed immune system to combat the seasonal viruses and bacteria—the *real* microscopic enemies—that I was encountering in my day-to-day life. My immune system was reacting to the wrong triggers, mistaking harmless substances for harmful ones, functioning so diligently that it was making me sick in the process.

It turns out that I am my father's daughter after all—we share a similar hypersensitive immune system—though I still don't know if I'm allergic to bees (but more on that a bit later). Over the ensuing months, as I slowly came to terms with the continuing mysteries and frustrations of my allergies and started to think of myself as an allergy patient, I took some cold comfort in the fact that at least I was not alone. Far from it. Once I revealed my own surprise diagnosis, people began talking to me about their own food, skin, or respiratory allergies. It suddenly seemed to me as if *everyone* I knew had some type of allergic condition; they just hadn't been openly discussing them. And that's when I realized that allergy was a much larger problem than I had ever imagined.

Nut allergies. Hay fever. Asthma. Eczema. Either you have a frustrating allergy or allergy-related condition, or you know someone who does. The latest statistics on allergies are sobering. Over the last decade, the number of adults and children diagnosed with mild, moderate, or severe allergies has been steadily increasing with each passing year. Billions of people worldwide, an estimated 30–40 percent of the general global population, currently have some form of allergic disease, and millions have one severe enough to actively endanger their health. But allergies don't have to be deadly to impact your whole life. People with mild, moderate, and severe—but not deadly—allergic immune responses spend an inordinate amount of time, money, and focus on their conditions. Allergies can be a burden, even when they aren't life-threatening. But because allergies don't normally kill people, as a society we have a tendency not to take them very seriously. We joke about someone's gluten intolerance or hay fever without thinking twice about how a person with those conditions might actually feel. The quality of life of someone with an active allergy is typically lower than someone without one. Their anxiety and stress levels are higher. They feel fatigued more often. Their ability to concentrate and their energy levels go down.

Maybe you already know what having an allergy is like because you have one. There's also a good chance that you've downplayed your own allergy because you've gotten used to the feeling of it. In other words, you've stopped expecting to feel "great" and have settled for feeling "okay" on most days of your life. But even then, even when an allergy sufferer has found ways to cope with their condition, there are times when it is harder to ignore. A bad pollen day. A new patch of red, itchy skin. A potluck dinner party. Allergy sufferers know what often remains hidden to the allergy-free—that our bodies are constantly bumping up against the billions of invisible particles, microbes, chemicals, and proteins that constitute the space and objects around us. Our immune cells make snap decisions, either to accept or to reject the things we encounter, countless times each day for our entire lives. Our immune systems decide, in essence, what can become a part of us (food), what can coexist with us (some bacteria, viruses, and parasites), what we can tolerate or ignore and what we cannot.

It's clear that our human immune system is becoming ever more sensitive to the panoply of natural and man-made allergens that we come into contact with on a daily basis. The problem is that immunologists working to understand the biological processes involved in allergic reactions aren't entirely sure why. Worsening food, skin, insect, drug, and respiratory allergies remain some of the most pressing medical mysteries of the twenty-first century. Why are we all so irritated?

———

After my own diagnosis, I went in search of more information on allergy. I wanted answers to a series of escalating questions that began with the very personal and cascaded into a set of larger historical, economic, social, political, and philosophical questions.

- How long have allergies been around? Are they an ancient problem or relatively new?
- Are allergies getting worse? If so, what might be causing it?
- Are allergies genetic, environmental, or man-made?
- What can we do about them? Can we "fix" our allergies?

After a few weeks of researching, I couldn't find any satisfactory and easily accessible answers. These questions turned into a personal and scientific journey to diagnose the problem of allergy in the twenty-first century. This book is a record of that journey, a holistic examination of the phenomenon of allergies from their first modern medical description in 1819 to the recent development of biologics for their treatment and immunotherapies for their prevention.

What you are about to read is an attempt to tell the whole story of allergies in the twenty-first century: what they are, why we have them, why they've been getting steadily worse globally, and what that might mean about the fate of humanity in a rapidly changing world. It interweaves the latest scientific research, the history of allergies, and the personal narratives of patients and doctors coping with allergies to explore our complicated connections to our environments.

First, we'll tackle the shifting definition of what an allergy is—and isn't. As our scientific knowledge related to immunology—the study of immune system function in all species—has deepened and progressed, so, too, has our understanding of what falls under the category of "allergy" or allergic-type immune responses. As we'll discover, allergies are not so easy to categorize, diagnose, and count. The best statistics we have are estimates based on insurance claims, surveys, and hospital admissions. But any way we do the math, the sheer number of allergic individuals is growing with each passing year—and with no end in sight.

Once we've learned the basics about allergies, we'll explore the various theories about their causation. Depending on how you define allergic immune responses, they are either very old—the ancient Egyptian king Menes is believed to have died from a bee or wasp sting—or very new. The first clinical description of an allergic response, an analysis of a case of hay fever, was penned just over two hundred years ago, and evidence suggests that respiratory allergies were not widespread until at least the beginning of the Industrial Revolution. Ideas about why rates of allergics have been steadily rising ever since are complex and heavily debated. If you want easy answers, you will not find them here. But you will learn what the likeliest combination of culprits are.

And, finally, we'll take a look at what treatments we have for allergy and what the future of allergy medicine is likely to be. Not much about allergy treatment has changed in the last two centuries, but a new class of biological drugs on the horizon might provide a glimmer of hope for better, and more consistent, relief of our worst symptoms. At the same time, new scientific understandings of our allergic immune responses might lead to better regulations and social policies. In the end, understanding what is irritating us and why, then and now, might help us to cooperate in order to craft better environments in the future—ones in which we can all breathe easier.

———

My father on duty in Vietnam. [AUTHOR'S PHOTO]

This book is dedicated to my father. My dad was an avid reader and a lifelong learner. Although he never finished his first year of college, he was a natural autodidact and enjoyed discovering new facts about the world until the day he died. In that way, too, I am firmly his daughter. I inherited not only his allergic tendencies but his curiosity and his constant quest for the truth—no matter how complicated and opaque that truth turned out to be. I think he would be entertained, enlightened, and fascinated by the story of allergy told within these pages. And whether or not you, my dear reader, have an allergy yourself or you love someone who does, I hope that by this book's end you not only have a better understanding of allergies but also have developed a few new questions about our incredible immune system and its complex relationship to our shared environments. Thank you for going on this journey with me. Let's get started.

PART ONE

Diagnosis

∘ ∘ ∘

The first step in our quest to better understand allergy in the twenty-first century is to survey all our current symptoms. In the next three chapters, we'll take a closer look at the problem of allergy today by analyzing the latest statistics and hearing from individual allergy sufferers about what it's like to have hay fever, allergic asthma, allergic dermatitis or eczema, food allergy, drug allergy, or insect allergy. To complicate things, it's not always easy to diagnose an allergy or to officially differentiate it from an intolerance or a sensitivity. Our immune system function is complex, and allergy is on a spectrum of possible immune responses ranging from full-blown allergic response to mild or moderate irritation to complete tolerance. To better understand what an allergy is and what it isn't, we'll explore the history of the immune system and how allergy fits into it.

CHAPTER 1

What Allergy Is (and Isn't)

————————

Before I began researching this book, I had no idea just how massive the problem of allergy truly is. Approximately 40 percent of the entire human population already has some form of allergic condition.[1] And by 2030, experts estimate that statistic will increase to 50 percent. But before we can dive more deeply into what these numbers might mean, and why allergies are projected to rise over the next few decades, we need to answer a more basic and fundamental question: *What exactly is an allergy?*

When I first started talking with scientists and allergists, I assumed I knew what an allergy was. If someone had quizzed me, I would have said, confidently, that an allergy was a negative bodily response to something a person had eaten, touched, or inhaled. If pressed for more details, I probably would have trotted out what I had learned long ago from an introductory biology course—that the human immune system is similar to a defense system. It reacts to foreign substances, such as viruses, bacteria, and parasites, and helps to protect us against infection. But in people with allergies, that same immune system is triggered by something in the environment—like pollen or milk or nickel in metal jewelry—that is harmless to nonallergic people. I would have listed sneezing, runny or stuffy nose, coughing, rashes, redness, hives, swelling, and difficulty breathing as possible symptoms.

Whenever I ask normal people (i.e., not scientists or biomedical experts) to explain what an allergy is, I usually hear something similar to my own initial definition. People of all ages and backgrounds tend to think of allergy and allergens as, as one young nonallergic man described them to me: "Some sort of imbalance with whatever is entering your system. It just doesn't mesh well with whatever is in your body and it causes your body to try to get rid of it." Another man described allergy as the body being "self-destructive" when it doesn't know how to handle something like pollen or a particular food. In one memorable interview, a man with several allergies who had grown up in Chihuahua, Mexico, near the Texas border, suggested that his body is in a constant defense mode—but sees this as primarily positive. He thinks of himself as well defended and described his body as more "careful" and alert than the bodies of nonallergic people. These are all more or less accurate depictions of allergic-type immune responses and they work well enough . . . until they don't.

Even people who have allergies don't always understand what, in exact terms, they are or how to distinguish them from nonallergic conditions with similar symptoms.

Take "Chrissie,"[2] for example, one of the first allergy patients interviewed for this book. By the time we spoke, Chrissie had been coping with respiratory allergy symptoms, hives, sporadic swelling of her eyes, and frequent stomach issues for years. She had been diagnosed with hay fever, or seasonal allergic rhinoconjunctivitis, and occasionally visited an ear, nose, and throat specialist (ENT) for treatment when her symptoms changed or worsened. She also experienced gastrointestinal symptoms and skin rashes if she accidentally consumed milk or gluten. Years ago, Chrissie went to see an allergist and was tested for reactions to the most common allergens. Her skin was completely nonreactive to all food allergens, and the allergist told her that it was extremely unlikely that the symptoms she experienced were due to a food allergy. Chrissie's ENT has repeatedly encouraged her to get retested, but she hasn't; instead, she goes online to research her symptoms and crowdsource possible remedies.

When asked to define what an allergy is, Chrissie said that it is what happens when the body can't handle something, especially if the body has come into contact with something too often or in too great a quantity. Over time and with repeated exposure, she explained, the body ceases to be able to process those things, giving rise to symptoms like her own. She doesn't believe the results of her skin tests for food allergens and insists that she has a food allergy; since wheat and milk are ingredients in most foods, she posits that her body has learned to reject them over decades of consuming them.

I am beginning this chapter with Chrissie's story—her misconception of what an allergy is and isn't, and her palpable confusion and frustration—to illustrate what we typically get right about allergy as well as what we typically get wrong. When it comes to her respiratory allergies, Chrissie is correct in thinking that her body is responding to something that it has had repeated exposure to, but she is wrong about her body being unable to process pollen. (As we'll soon see, it's more that her body isn't able to tolerate it or ignore it.) Chrissie likely doesn't have a true food allergy, despite having very real symptoms, because she doesn't show any sensitization to milk or gluten (as evidenced by the results of her skin-prick test). In other words, her immune system is likely *not* reacting to the foods she's ingesting. Her immune system *is* reacting to pollen, however, which causes her hay fever. What Chrissie is really confused about, then, is the difference between an intolerance (in this case, to certain foods, possibly caused by another condition like irritable bowel syndrome or a lack of the enzyme lactase that aids in breaking down the lactose in milk products) and an allergic response (to airborne allergens). And who could blame her? Even as a medical anthropologist with a decent understanding of immunology, I had to discover some of these distinctions the hard way.

The deeper I waded into the scientific literature on allergy and the more conversations I had with allergists and immunologists, the murkier the definitional waters got. To my initial surprise and frustration, the more I learned about how the intricacies of our immune system function, the harder, and not easier, it was to understand allergy. It turns out that what we commonly refer to as "allergy" is actually a grab bag of various conditions. The one thing they all have in common is this: They all involve a hypersensitive immune system reaction to an otherwise innocuous substance—an allergen—that doesn't typically produce any immune response in nonallergic people. The symptoms of an allergy vary depending on how the allergen enters the body (via the skin, airway, or intestinal tract), the individual genetics of the person, and the many different "allergic pathways" the allergen can trigger.

So, then, what is an allergy? It's a harmful immune-mediated hypersensitivity reaction to a harmless antigen, which is defined as any toxin or foreign substance that activates an immune response. That's the technical scientific definition, but it likely doesn't mean much to you—yet. To fully comprehend what an allergy is, we have to understand how the definition of the term itself shifted and changed over the past century. The concept of allergy is just over a century old, born out of early studies of the function of the mammalian immune system.

In the end, and as you'll soon see, I learned that an allergy is perhaps best defined by what biological processes it sets in motion.

THE EVOLUTION OF A HERETICAL IDEA:
A SHORT HISTORY OF ALLERGY

Before we dive into the complicated, intertwined history of allergy and our understanding of the immune system, it's important here at the outset to underline that an allergy really isn't a "thing" at all, at least not in the ways we're used to thinking about other concrete things that exist in the world—like tables or viruses or cats. Instead, it's better to think of an allergy as a complex biological process involving many different, intersecting components of our immune system. Allergy is more about the actions our immune cells decide to take than it is about the symptoms we might experience because of those actions. The story of how our knowledge of immunity evolved and made the discovery of allergic reactions possible begins in earnest at the turn of the last century.

Our ideas about the immune system, both past and present, owe a lot to our earliest understanding of microbes. By the late 1800s, famous scientists like Louis Pasteur, Joseph Lister, and Robert Koch were busy conducting experiments to definitively prove that living organisms we cannot see—such as anthrax, tuberculosis, and cholera bacteria—can make us sick, infect wounds, and rot food. This new understanding of contagion and the workings of microorganisms—typically referred to as the "germ theory" of disease—gave birth to the modern medical concept of immunity, or an organism's ability to stave off illness.

To be immune is to be protected from or defended against infection from any particular external organism. The biological mechanisms behind immunity became the focal point of scientific research in germ theory throughout the late nineteenth and early twentieth centuries. By the 1900s, scientists were focused on understanding the basic biological mechanisms that produced either immunity or disease in an individual animal after it was exposed to a disease-causing organism like the anthrax bacillus. The ultimate goal of these early immunologists was to learn how to induce immunity. At the time, vaccines and serums containing small amounts of altered microbes and disease-fighting antibodies were already being used in medical clinics and hospitals, either to prevent or to treat commonplace maladies, such as smallpox, diphtheria, or tetanus, but the process by which they worked remained almost completely shrouded in mystery.

Spurred on by the success of these early vaccines and serums, scientists and doctors firmly believed that it might be possible to produce immunity to *all* infectious human diseases and toxins. All that was needed, they thought, was a better working understanding of how animals developed immunity in the first place. Global efforts to produce immunity and treat a variety of diseases provided the backdrop for the accidental discovery of allergy.

The term "allergy"—meaning "different activity" from its combined Greek roots *allos* and *ergon*—was first coined by Clemens von Pirquet, a doctor working in a pediatric clinic in Vienna, Austria, at the turn of the last century. Pirquet and his colleague Béla Schick noticed that some children given smallpox vaccines made of horse serum (a common medical practice at the time) would react poorly to a second dose, developing a rash at the injection site, itching or inflamed skin, and a fever. Surmising that something in the serum itself was causing these negative biological reactions, the duo started to methodically observe their patients after repeated injections of smallpox vaccine.

Initially, Pirquet used "allergy" to indicate *any* altered biological state, good or bad, that had been triggered by an exposure to a foreign substance—in this case, the serum.[3] For Pirquet, a negative altered state or reaction might refer to the rash or fever produced by injections of a vaccine; a positive altered state or reaction might refer instead to the development of immunity produced by the same injections. Allergy, in its original framing, included *both* immunity and hypersensitivity. It was a neutral term meant only to indicate that something had induced a change in a patient's biological state of being.

In 1906, when Pirquet invented the term "allergy," immunity itself was still a fairly new and extremely limited concept, used only to refer to the body's natural defenses against disease.[4] As an idea, immunity had its beginnings in the realm of politics, not medicine, and was originally used to refer to an exemption from legal punishment or obligation.[5] Early scientists borrowed the term "immunity" and altered its meaning—but only slightly. In the realm of medicine, immunity referred to a natural exemption from infectious disease and indicated the status of being wholly protected against the "punishment" of illness, and perhaps death. The "immune system" itself was named for this version of immunity and, at that point, was essentially a working theory, meant to allude to any biological processes going on inside the body that were responsible for conferring it. At that point, the immune system's sole function was thought to be defense—*and only defense.* Early clinicians like Pirquet and Schick, who observed their patients reacting negatively to the same substances that should have produced immunity, thought that what they were witnessing had to be a phase in the systematic development of defense against that substance. They saw rashes, fevers, and itching at injection sites as evidence that the vaccines or serums were working; they were causing their patients' defense mechanisms to kick in.

But what if, as Pirquet and Schick began to realize, the immune system could make a mistake? What if our immune systems could make us sick as well as protect us? What if it wasn't just bacteria or toxins that could cause illness, but the so-called immune system itself?

This idea was revolutionary, heretical, and—at least initially— reviled and rejected. It was inconceivable for early scientists working in the field of immunology to accept that a person's immune system could play a role in causing them harm. The human body's production of antibodies[6]—the immune system's ability to create specialized cells that work to counteract harmful invading organisms—was thought to be purely beneficial. The realization that the same immune system responsible for fighting off bacteria might be the root cause of hypersensitive reactions to things like horse serum and pollen flew in the face of decades of work. Pirquet's theory of allergy directly challenged a fundamental tenet of the new field of immunology and, as a result, was largely dismissed. It would take more than a decade for scientists to realize not only that it was basically correct, but that it could be medically useful.

As more and more clinical and laboratory evidence piled up, scientists slowly began to realize that Pirquet's description of allergic reactions was far more prevalent than they had anticipated. At the same time, physicians started to recognize that so-called allergic reactions could also more easily explain many of the chronic illnesses—periodic asthma, seasonal hay fever, recurring hives—they were used to seeing in their clinics. As the years rolled by, the concept was more widely adopted as doctors working to treat otherwise perplexing maladies began to see "allergy" as a way of giving these patients a diagnosis that could at least partially explain what they were experiencing. Over time, the definition of "allergy" shifted to refer almost exclusively to these more troublesome and harmful immune system responses, so-called *overreactions* to otherwise harmless agents.[7]

By the mid- to late 1920s, the nascent field of allergy was just beginning to professionalize as a subfield of immunology.[8] As a term, it was regularly being used interchangeably with words like "sensitivity," "hypersensitivity," and "hyper-irritability" to indicate any overreactive immune response to an otherwise "harmless" substance. One of the leading allergists of the period, Warren T. Vaughan, defined allergy as a "hyper-irritability or instability of a portion of the nervous system."[9] As both a physician and an avid scientific researcher, Vaughan was puzzled by the idiosyncrasy of his individual patients' reactions to allergens. There was no pattern that made sense to him and no explanation for why, when controlling for all other variables, two people might react so differently to the exact same exposure to an allergen. Even more confusing, *the same patient* might respond differently to *the same stimulus* on different occasions or at different times of the same day. It was as if allergic reactions followed no biological rules whatsoever—at least none that Vaughan could easily discern.

By 1930, Vaughan had surmised that the overall purpose of the mammalian immune system was to maintain some kind of "equilibrium" or balance between the organism and its environment. An allergic person's symptoms, then, were simply signs of a temporary or chronic imbalance between that person and the rest of the biological world. Vaughan thought—correctly, as it would turn out—that an allergic reaction began on the cellular level rather than on the humoral, or whole body, level. When an allergic person's cells encountered a foreign substance or experienced an exogenous, or outside, shock, they overreacted, throwing their own biological systems off balance, either temporarily or chronically. The goal of the allergist was to help bring their patient back into a "balanced allergic state"— and then keep them there. The delicate equilibrium between "normal" and "allergic" states of being, at least according to Vaughan, could be upset by any stressor in the patient's life—a bad respiratory infection, a sudden change in temperature, a shift in hormones, or a generalized increase in the patient's level of anxiety.

Other early allergists defined the affliction in a similar way and posited many of the same causes for its onset in their patients. In the United Kingdom, Dr. George W. Bray defined allergy as "a state of exaggerated susceptibility to various foreign substances or physical agents"[10] that are otherwise harmless. For Bray, both anaphylaxis and allergy were best viewed as "accidents in the course of defense." Dr. William S. Thomas defined allergy as an "altered reaction"[11] and questioned the relationship of allergy to the development of immunity after repeated bacterial or viral infections (itself a faint echo of Pirquet's original thesis that immunity and hypersensitivity were related).[12] By the time of Thomas's writing in the 1930s, allergy researchers had already noted that asthma was often precipitated by a bacterial infection of the lungs and had started to surmise that there was a connection between a patient's prior respiratory illnesses and the development of an allergy. In a publication meant for medical practitioners, Dr. G. H. Oriel argued that there were only three possible states of immune system function: normal (neither allergic nor immune—neutral), sensitization (allergy), and immunity.[13] By the end of the 1930s, the term "allergy" had firmly gone from being a more neutral connotation of *any* biological change induced by an outside stimulus to a wholly *negative* description of a much more limited set of physical reactions to the introduction of any outside substance into the body. As a medical term, "allergy" had definitively shifted "to represent the dark side of immunity" by the 1940s.[14]

This reputation of allergy as "the dark side of immunity" was bolstered in the late 1950s when the famed immunologist Frank Macfarlane Burnet discovered that certain diseases such as lupus and rheumatoid arthritis were ultimately the result of the human immune system's inability to tell "good" cells from "bad" cells, or "self" from "nonself." Autoimmunity—when the body attacks itself—took center stage in immunological research after Burnet realized that the immune system's main function was *not* defending the body from infectious invaders but recognizing the body's own cells from everything else. After coming into contact with something from its immediate environment, the immune system could either choose to tolerate the foreign or "nonself" substance (as it does with most proteins ingested as food) or to attack it (as it does with many viruses and bacteria). In someone with an autoimmune disorder, the immune system makes a fundamental error, confusing the body's own cells for foreign cells, and becomes hypersensitive—or overreacts—to them. In essence, the immune system triggers a response to the body's own tissues.

Burnet's insights about autoimmunity would provide the groundwork for further scientific research on immune function for much of the twentieth century, as the field of immunology became more and more focused on understanding the development of immune tolerance rather than defense. Today, allergy and autoimmunity are largely seen as variations on the same theme rather than as entirely different problems. Both highlight how the biological mechanisms behind our immunity to disease and our tolerance of natural and man-made substances can go awry. In the twenty-first century, Pirquet's original suggestion that our immune systems could just as easily harm us as protect us is no longer a heresy but rather a commonplace understanding of our overall immune function—and dysfunction.

More recent work in immunology has shifted once again, this time away from Burnet's self/nonself paradigm toward a model that reflects our current understanding of how our own cells interact with the trillions of nonhuman cells, particles, and chemicals in our intestinal tracts, in our nasal cavities, and on our skin. How do our bodies decide what things to tolerate and what things to fight? In other words, our immune cells need to determine when our bodies are in harm's way from something in our environments and when they aren't. How they do this, however, remains a mystery. Dr. Pamela Guerrerio, one of the top food allergy researchers and clinicians working at the National Institutes of Health (NIH), explained that "we still don't understand the mechanisms behind immune tolerance, to be honest, or why we tolerate some things, but not others." Dr. Avery August, an immunologist at Cornell University, told me that debate still rages over what the ultimate function of our immune cells might be. While it is clear that they provide protection against infection, August prefers to think of immune cells as the "curators" of our body, constantly sensing everything we encounter and making millions of microdecisions about which things should become part of the human body or coexist with us and which things should not. The only thing we seem to know for certain about our immune system is that, as it becomes more irritated in the twenty-first century, it is less and less able to tolerate even some of the stuff that is "good" for us in our environment.

HOW ALLERGY IS DEFINED TODAY

As we've seen, defining exactly what an allergy is has been a problem since its inception. In 1931, the renowned allergist Dr. Arthur Coca argued that using "allergy" as a medical term wasn't particularly useful because clinicians and other nonspecialists tended to use it to mean anything.[15] It had become a "grab bag" diagnosis, used to assuage patients when all other diagnoses and treatments had failed.

Allergists and scientists I talk to often echo Coca's lament: They tell me that one of the toughest and most consistent problems they face is the general misconception about what an allergy really is. In conversations with me they repeatedly argued that the public often uses the term indiscriminately to describe almost any uncomfortable set of symptoms they might experience. If people have frequent indigestion or experience pain after a meal, they may attribute it to an allergic reaction to something they've eaten—like dairy—even though they never visit an allergist to confirm or disprove their suspicions.

Over the past one hundred years, allergy has become a popular and widely used medical concept, but one that isn't always applied appropriately or effectively. Allergists and immunologists want everyone to understand that an allergy is not the same as a sensitivity, an intolerance, or an autoimmune disorder. The main difference lies in the biological processes or immune mechanisms that are activated.

A Quick Primer on Our Immune Systems

The first thing you should know about the human immune system is that it is actually made up of two different systems that work in concert. The innate immune system, fully functional from birth, is a brute-force, first line of defense against foreign invaders like pathogens. Because it reacts in the same way no matter what foreign object it encounters, it is sometimes referred to as the "nonspecific" system. Your skin and mucous membranes—or the outside and inside linings of your body—are part of the innate immune system. If something gets past these barriers, then the innate immune system can activate inflammation to ward off microscopic invaders. Mast cells and basophils (which we've already seen at work during anaphylaxis) are involved in this process. Special scavenger immune cells called phagocytes can engulf or "swallow" bacteria, killing them, and natural killer cells can use toxins to destroy any cells that have already been infected by a virus. These different components of the innate immune system are often enough to ward off infection.

The adaptive immune system kicks in if the innate immune system hasn't been able to deal with the threat. In this book, we're going to pay the most attention to the adaptive immune system because it's behind our hypersensitivity reactions (which include autoimmunity and allergy). As a second line of defense, the adaptive immune system is a "specific" system because it is capable of remembering the specific things it encounters and, upon a subsequent exposure, reacts accordingly. T lymphocytes, a type of white blood cell produced in our bone marrow, have detection features on their surface that attach to foreign invaders like germ cells in our bodies. After coming into contact with a specific foreign invader, some of these T cells can become "memory" T cells. The next time they encounter a similar organism, they can activate the adaptive immune system much more quickly. B lymphocytes, another type of white blood cell produced in our bone marrow, are activated by T cells. B cells can quickly produce large amounts of antibodies and release them into the bloodstream to help fight off foreign cells. Antibodies are Y-shaped proteins that circulate throughout your blood and whose main function is to neutralize foreign substances like viruses and bacteria. Antibodies attach themselves to foreign microorganisms thereby preventing them from being able to attach to or penetrate our own cell walls. At the same time, antibodies can attach themselves to other immune cells and activate them, aiding in and promoting an overall immune system response. Antibodies are specific to the type of B cell that is producing them and the type of T cell that triggered the process, so they are "ready-made" to defend against a specific type of foreign material that has entered the body— one that the body has "remembered" from a prior encounter.

Our bodies produce five different types of antibodies: IgM, IgD, IgG, IgA, and IgE. We'll meet both IgG and IgE again, but it's IgE that will be the main focus for much of this book. While not all Type I hypersensitivities—also known as allergic immune responses—are mediated by IgE, most allergic responses typically involve the activation of IgE. In contrast, Type II and Type III hypersensitivity immune responses, which include immune conditions like Graves' disease and autoimmune disorders like lupus and rheumatoid arthritis, are mediated by IgG antibodies. For better or for worse, an IgE antibody response has become the main indicator of an allergic-type immune response and is synonymous with allergy. A genetic predisposition to IgE sensitization to allergens in the environment is called atopy. So (and this will be important later), atopy is different from allergy because *while you can have an allergy without an IgE response, you cannot have an atopic reaction without IgE.*

This connection between IgE and atopy was an important discovery that led to major innovations in research on allergic responses and their treatment. However, it also causes confusion when it comes to parsing the differences between allergy, atopy, and things like intolerances or sensitivities (as we'll see in chapter 2 on diagnosis). Because of IgE's central importance as a marker of allergic response, I want to pause here and take a quick detour to explore the discovery of the antibody itself.

The Discovery of IgE

As early as 1906, when he coined the term "allergy," Clemens von Pirquet posited (correctly, as it turns out) that allergens were activating an antibody response in his patients. In 1919, Dr. Maximilian

Ramirez reported that one of his patients had developed an allergy to horse dander after receiving a blood transfusion from an allergic donor.[16] This was proof of Pirquet's guess that something in the blood could transfer an allergic sensitivity, possibly a new type of antibody itself. Then, in the 1920s, Dr. Carl Prausnitz, a doctor working in Germany who was allergic to rye grass, attempted to transfer his own natural allergic sensitivities to pollen to his assistant Heinz Küstner, who was allergic to cooked fish, and vice versa.

By this time, it was clear that skin-prick tests worked to elucidate sensitivities to different allergens (more on this in chapter 2), but the biological mechanism behind those reactions remained a puzzle. After transferring Küstner's blood serum into his own arm, Prausnitz developed a wheal response to fish allergen during a subsequent skin-prick test. Despite several attempts using different blood serums derived from patients with more severe allergic reactions to rye pollen, Küstner never developed a positive skin reaction to pollen himself. Yet Prausnitz's own positive skin reaction to fish proteins had proven that allergic sensitivity could be transferred via blood serum infusions. The pair's research led to the development of the Prausnitz–Küstner reaction, or the P-K test for allergic sensitization, and was widely used by allergy researchers for decades. But even though the P-K test was useful for immunological research on hypersensitivity, the biological mechanisms behind it were clouded. After decades of scientific investigation, immunologists thought it was likely that *some* type of antibody was responsible for inducing sensitivity during the P-K test, yet most of the known antibodies had been ruled out as the culprit.

The stage is now set for the discovery of IgE.

In the late 1960s, two Japanese researchers decided to study P-K activity in the serum of patients allergic to pollen. At the time, immunologists were suspicious that the reactivity of the skin during P-K tests might be related to the action of the antibody IgA. But after several experiments, Drs. Kimishige and Teruko Ishizaka determined that the biological activity they were witnessing could not be caused by any of the known antibodies—IgM, IgA, IgG, or IgD. The Ishizakas' work revealed that a new antibody, which they named IgE, was binding to mast cells and basophils, helping to drive the allergic response. The Ishizakas' subsequent careful scientific research on IgE function definitively proved it was the antibody involved in most sensitivities or immune overreactions to otherwise harmless antigens or allergens.

An antigen is any substance that initiates an immune system response; an allergen is a type of antigen that triggers an IgE antibody immune system response. In this sort of response, your body's immune cells trigger what's called the Type I allergic "pathway" (which is why researchers refer to allergy as a Type I immune response). Some of your immune cells—a subset of the white bloods cells known as CD4⁺ T cells known as T helper type 2 (Th2)—signal B cells, another type of white blood cells, to produce IgE antibodies. Of the five types of antibodies found in mammals, IgE is the only one known to regularly bind to allergens to begin an immune response. And unlike the other antibodies, which are found in the blood, lymph, salivary, and nasal fluids, IgE antibodies are localized in our tissues, where they are tightly bound to the surface of our mast cells (some of your immune system's first responders). IgE antibodies are primarily responsible for binding to parasites like intestinal worms, but in an allergic reaction, they trigger your mast cells and basophils (the other first responders) to release histamine and other compounds that then cause inflammation and all of the symptoms you typically associate with your allergy. Atopic, or allergy-prone, people tend to have not only higher levels of IgE but also more receptors for IgE on their mast cells, which is likely part of the reason they are more sensitive to things in their environments in the first place and tend to develop allergic responses to multiple allergens. However, someone who is non-atopic—that is, a person who doesn't have a biological tendency toward sensitivity (we'll look at the difference much more closely in chapter 4)—can still develop an allergic response to bee venom or penicillin, for example, if repeatedly exposed.

The discovery of IgE's role in allergy paved the way for more scientific research on the specific mechanisms or "immune pathways" that a person's body might take to spiral into a hyperactive immune response. Scientists and clinicians today distinguish between IgE-mediated allergies (e.g., allergic rhinitis, food allergy, atopic eczema) and non-IgE-mediated allergies (e.g., drug allergy, serum sickness). But in essence, and for all practical purposes, the term "allergy" in the twenty-first century has come to mean *any negative immune reaction driven by IgE antibodies*. The presence of IgE in response to an exposure to an antigen has become the standard of measurement and confirmation for what is known as a Type I hypersensitivity or "allergy."

The Problem with Relying Solely on IgE to Define Type I Hypersensitivity

Using the presence or absence of IgE antibodies alone to categorize an allergy quickly becomes problematic if a patient has low levels of the antibody to begin with. It might also exclude other allergic conditions like eosinophilic esophagitis (EoE) and nonallergic eczema because they are thought to be non-IgE mediated. In fact, serum sickness, or the reaction that Clemens von Pirquet observed in his Kinderklinik and used to coin the term "allergy," falls into the category of non-IgE-mediated allergic disease. People who have asthma or atopic dermatitis and who do not make IgE in response to exposure to an allergen can be classified as having a "Type I allergic disease" because the same core physiological responses are involved, but not an "allergy" in the strictest definition of the term if we use IgE as the litmus test.

It is important to note that some of the experts I interviewed for this book were very comfortable with calling eczema or asthma an allergy; others were adamantly opposed to it. Some felt that the trigger of an asthma attack or an eczema outbreak mattered more than the response. For example, if someone has an asthma attack during strenuous exercise, it would not be accurate to lump that person in with people whose attacks are triggered by *allergens* like the level of grass pollen in the air. Those who argue that the underlying biological mechanisms that drive the response in each case are the same—and that those biological pathways matter more than the triggers—are more at ease with saying that asthma and eczema are allergic disorders. In many ways, the current debates over what does and does not fall into the category of an allergy are a continuation of the debates over the meaning of the term itself in the early twentieth century. If you're still confused about what an allergy *really* is and how we define it, you're not alone.

Today's allergists are divided over how they distinguish between these conditions and the exact meaning of the term "allergy." Many doctors I interviewed expressed a desire for more precise definitions or new terminology.

Dr. Hugh A. Sampson, a world-renowned allergist with forty years of experience in the field, says that an allergic reaction is unique to each person and can express itself differently through time. In young children, an allergic reaction typically affects the skin and the gut. A baby experiencing a reaction to a food will get a skin rash or throw up. But as she ages, the target organs may change, and she may start experiencing allergic reactions as asthmatic episodes or wheezing. "Allergy refers to a common underlying immune mechanism," Sampson explains. "It's just that each response can target a different organ."

Dr. Gurjit "Neeru" Khurana Hershey, a pediatrician, an endowed professor, and the director of the Division of Asthma Research at Cincinnati Children's Hospital, defines allergic disease as a "whole body or systemic disorder." In some people, the allergic response will be targeted in one area such as the respiratory tract; in others, it will manifest itself in multiple areas, as in a person who suffers not only from asthma but also from eczema and food allergy. But in either case, it's still a *systemic* disease. Inflammation is the core problem of *all* allergic disorders—the commonality that unites all the conditions under one umbrella term. The puzzle, as Khurana Hershey sees it, is to understand why that response remains localized in some patients and distributed in others.

The branch chief of Allergy, Asthma, and Airway Biology at the NIH, Dr. Alkis Togias, describes allergy as a syndrome, or a group of symptoms that commonly occur together and have at their root the same underlying cause. From his vantage point, asthma, hay fever, eczema, and food allergy are *not* separate problems.

"We're really dealing with one syndrome which expresses itself in different parts of the body," Togias explains to me. He blames some of the confusion over what is and is not an allergic condition on the hyper-specialization of medicine during the past few decades. Pulmonologists primarily deal with the lungs, so they will diagnose asthma. But they won't necessarily notice, or care, if a patient also has eczema or a food allergy—even though as an atopic person they often do. Togias argues that we end up treating these as separate conditions in the same patient, despite the fact that they are expressions of the same syndrome. In other words, not everyone who has an allergic disease will be diagnosed or treated by an allergist; nor will they think of their separate allergic conditions as part of the same underlying immune malfunction.

Dr. Donald Leung, a prominent allergist and immunologist at National Jewish Health in Denver, argues that terminology is a big part of the confusion. Allergic conditions are often categorized more by their symptoms than by their biology—"wheezing" for asthma, "itching" for atopic dermatitis. He thinks that "atopy" is a better term than "allergy," since atopy literally means "out of place." The reaction of a person's skin, gut, or nasal cells to an allergen are "out of place"—they are overexuberant responses to common and otherwise harmless stimuli in their environment. Ulimately, then, his definition is all about the underlying reaction of our immune system, and not just our symptoms or the results of our allergy tests.

ALLERGY MADE (SORT OF) SIMPLE

So where does all this leave people like Chrissie with her mistrust of her negative food allergy tests; or you, dear reader, who may have allergy symptoms but have never seen an allergist; or my dad, with his non-IgE-mediated, but deadly, reaction; or me, with clinical symptoms of respiratory allergies but no evidence of IgE response during skin or blood tests (we'll examine this mystery a bit more in chapter 2)? In other words, how should nonexperts understand allergy?

The remainder of this book will use the definition of Type I hypersensitivity as a starting point. For simplicity's sake, I'll be using a basic rubric for determining what an allergy is—and what it isn't. If your body's immune system responds to an exposure to an antigen or allergen, then you have an *allergy*. Typically, that will usually mean that you also have an IgE response, but not necessarily. What's more important is that your immune system has a hyperactive response to an otherwise harmless substance. If you have symptoms similar to a food allergy, but they are being caused by some bodily system, condition, or mechanism other than the immune system itself, then you have an *intolerance* (which, to be perfectly clear, is not an allergy). If you develop a localized wheal reaction to a skin-prick test (which we'll explore more in chapter 2) but don't experience allergic symptoms when exposed to that allergen, then you have a *sensitivity,* not an allergy.

Hopefully, this brief synopsis of all the science we've covered so far makes the definition of an allergy easier to understand. But if you still think some of this is a bit confusing and convoluted, don't worry— it is. In fact, it's often not even that clear to clinicians trying to accurately identify an allergic condition in the first place. This diagnostic messiness is the subject of the next chapter.

CHAPTER 2

How Allergy Diagnosis Works (or Doesn't)

———

A TYPICAL ATYPICAL DIAGNOSIS

"You're almost like a detective, in a way," Dr. Purvi Parikh says. We are sitting in her office talking about what it's like to be an allergist in the twenty-first century. It's quiet, just after business hours, and the waiting room is unusually dark and empty. Parikh has been an allergist for more than a decade and is a clinical assistant professor in the Department of Pediatrics at the Grossman School of Medicine at New York University. She specializes in asthma care and does research on asthma in children, but she treats patients with all forms of allergies at her Midtown practice. Parikh assures me that if I were visiting in the summertime instead of the dead of winter, the waiting room would still be packed at this hour with patients seeking help for the worst of their seasonal respiratory symptoms. But it's January, so we have time for a more in-depth talk.

Parikh admits that she really likes helping people figure out their allergies; it's what first drew her to the specialty. She was just out of medical school and a new attending physician when a male patient undergoing open-heart surgery went into shock on the operating table. No one could figure out why until Parikh realized that the patient could be having an allergic reaction to something. She ran some tests on a hunch. The results showed that the patient had a severe allergy to the antiseptic solution being used to prep him for surgery.

The patient had never had an allergic reaction before and had no idea he was allergic to anything. After Parikh's discovery, the man's surgical team switched antiseptics and successfully performed his surgery. That was the first time that Parikh felt the thrill of figuring out a difficult case and helping a patient get the treatment he so desperately needed. She was immediately hooked.

It's easy to see from her enthusiasm that Parikh loves her profession. But she warns that it can also be a challenging job—far more difficult than many nonexperts might surmise. Allergy is a medical subspecialty that relies on the clinician's experience and gut instincts as much as it does upon modern diagnostic tools and a patient's biomedical history. That's why Parikh likens her day-to-day to that of a detective. Diagnosing an allergy is never easy. It's always, in some fashion, like solving a medical mystery. People with milder symptoms or "hidden" allergies often just don't feel a hundred percent and know something isn't quite right—the ultimate reason *why* is what allergists must help them unravel.

Like Tolstoy's unhappy family, each allergy patient suffers in their own unique way. No two allergy cases are alike, and an official allergy diagnosis can take hours, days, weeks, months, or even years. That's because allergies are biologically complex, test results can be inconclusive, and most common symptoms mimic other diseases.

"It can be very gratifying to help a patient come to a diagnosis," Parikh tells me. That's when she turns her attention to me, seeing me not just as a scholar researching her field but as a person in need of her skills. She's baffled that I haven't seen an allergist myself, since I have symptoms and my father died of a bee sting. She fixes me with a friendly smile and says, "I really think you should make an appointment to see me. I think you should be tested. We should figure this out."

Like so many people with allergy symptoms, I had been hesitant to see an allergist. Since my symptoms were generally mild and readily controlled with over-the-counter antihistamines, it had been easy to delay seeking out more professional care. But I knew Parikh was right, so I took her advice—eventually.

———

By the time I found myself in her office again, an entire year had gone by, and my sinuses were really bothering me. As instructed when I made the appointment, I hadn't taken any antihistamines for a week. After a brief consultation, Parikh called her nurse into the exam room to perform a standard skin-prick test for reactivity to specific allergens and a quick breathing test to check to see if I might have mild asthma on top of any potential allergies.

The nurse was in her midforties like me, tall, friendly, and dressed in colorfully patterned scrubs. She escorted me to the end of a short hallway; there sat a spirometer, a special machine that measures the air pressure generated by my lungs. As I forcefully blew into a plastic mouthpiece connected to a tube, I watched as each out breath was measured on a graph on a computer screen in front of me. After three measurements, the nurse pronounced that I was firmly in the normal range: no asthma at all. I followed her back into the exam room and minutes later I'm wearing a paper gown covered in red lobsters, blue puffer fish, and yellow octopi. A large portion of Parikh's patients are children, and the brightly colored exam gown makes for a good distraction—even for an adult like me.

The nurse came back into the room carrying three small blue plastic trays. The trays contained white plastic applicators that look like eight-legged insects. Each leg ends in a sharp point that, when pressed lightly into the upper arm or back, slightly scratches the skin to deliver a trace amount of an allergen extract just under the first dermal layer. Allergists prefer doing the tests on the arm so that the patient can see the reactions for themselves, since seeing their skin reactions is often the patients' first step in understanding their reactivity. All told, I'm tested for reactions to more than fifty different allergens, including tree and grass pollens and common food allergens like egg and wheat. The test also includes a negative control that normal skin shouldn't react to (saline) and a positive control that normal skin should react to (histamine) to make sure the test is functioning properly and the results are accurate. Once the nurse had marked my arm with corresponding numbers so Parikh could easily read the results, she carefully pressed the applicators onto my forearms and upper arms, rocking them gently back and forth. I felt the plastic tips dig in. The nurse then exited the room, and I was left to stare at my skin for the next twenty minutes, which is the average amount of time that it takes for skin cells to react to every allergen.[1]

Immediately, I felt the histamine control starting to work. The skin underneath the minuscule scratch began to itch—mildly at first and then uncontrollably. I had a difficult time not scratching at it. I stared at my arm and watched a pink elevated wheal form where the histamine was, akin to a large mosquito bite. In a sensitive person, the skin immediately reacts to the allergen, producing an inflammatory response at the site of injection. This is referred to as "a wheal and flare reaction" by allergists. The release of histamine from the patient's mast cells is the main driver behind this reaction. Typically, a patient is considered positive for sensitivity if a wheal greater than 3 millimeters develops and flare diameter is greater than 10 millimeters. However, if the positive control produces a wheal and flare smaller than 3 millimeters, that might be used as a way to assess the other wheals. Any wheal size is considered evidence of allergic sensitization, though smaller wheals are considered less predictive of an actual allergy.[2] I watched for reactions at the other numbered sites, but all I saw were the leftover drops of allergen extract drying on my pale skin. After the allotted time, Parikh knocked on the door and peeked in. She scrutinized both arms and intoned "hmm" before informing me that my skin hadn't reacted to any of the allergens.

"That doesn't necessarily mean you're not allergic to any of these," she explained. "It just means that we have to dig deeper, pardon the pun."

What typically follows a failed skin-prick test is an intradermal test. Intradermal tests use a traditional syringe to deliver a small amount of allergen extract deeper into the skin. Parikh's nurse returned with a metal tray filled with twenty different syringes. She cleaned off both my upper arms with alcohol pads, erasing the pen marks and any remaining extract, before lightly pinching my skin to inject me. One by one, the needles punctured my skin. By the time the nurse finished, my skin looked angry. At the puncture sites, some small droplets of blood and raised bumps formed. I'm then left alone to wait for another twenty minutes. This time, as I stared at my arms, I thought about my father and one of my aunts on my mother's side who has severe allergies. I wondered how much of my immune response would mimic theirs—or not. But other than the puncture wounds from the needles and another itching patch of skin where the histamine was injected, nothing was happening.

After the allotted time expired, Parikh came back into the room, looked carefully at my arms, and sat down. "First," she said, "I want to stress that I believe you. I believe that you have clinical symptoms of allergy." She paused a moment, her wide, bright eyes looking directly into mine. "The problem is that your skin is one hundred percent nonreactive. It happens."

Parikh explained that in a small subset of patients with obvious respiratory allergy symptoms, the skin cells are far more tolerant of the allergen than the cells that line the sinuses. In other words, I may indeed have a legitimate, convincing case of seasonal hay fever or year-round respiratory allergies, but it will never show up on any skin test. The cells that line the skin and the cells that make up the mucous membranes might come into contact with the same allergen and respond very differently. But because she is thorough and I have decent health insurance and she likes to solve a good mystery, Parikh decided to order a serological allergy test. In a serological test, a patient's blood serum is mixed with allergens and checked for a resulting antibody response. If IgE, which, as you'll remember from chapter 1, is linked to atopy and is often predictive of an allergic reaction, is activated in response to the introduction of an allergen, then the patient is said to have a sensitivity to it (complicating matters, standard diagnostic tools only test for sensitivity and cannot always accurately predict whether or not a patient has or will ever develop an allergy).

Parikh filled out a form for the tests and I left her office to walk down the street to a nearby lab. After waiting nearly an hour, I had three vials of my blood drawn. The technician informed me that the whole process would take about a week.

I went home to wait for my results. But then a global pandemic interceded. It was late February 2020 and New York City was about to begin a lockdown in an attempt to slow the spread of SARS-CoV-2. It would be months before I got my antibody test results back, and even then the follow-up visit would be virtual. When Parikh and I finally spoke again that May, it was in the middle of a particularly bad spring pollen season, and I was coping with full-blown hay fever symptoms. My eyes itched and burned and sometimes spontaneously produced tears as if I were crying. My nose was permanently stuffed, despite my daily doses of allergy medicine. I was eager to learn which trees or grasses might be the cause of all this low-level discomfort.

"You're special!" Parikh announced at the start of our call, as if she were telling me that I'd won a coveted prize. "According to these results, your blood showed no reactions whatsoever. You're completely nonreactive. In fact, you had really low levels of IgE antibody across the board. If I were just looking at these test results, I'd say you weren't allergic to anything at all."

For a brief and silent moment, I felt slightly crazy. If each test that I'd taken—skin-prick, intradermal, blood antibody—was 100 percent negative, then did I really have an allergy at all? Or had I been imagining my itchy eyes and stuffy nose? What else could be causing the visible nasal irritation that an otolaryngologist diagnosed years ago and that I experience every spring, summer, and fall?

"I believe that you have clinical symptoms," Parikh said, as if reading my mind. "I absolutely think that you have an allergy. It's just that for some patients, their allergy is not IgE mediated and there is no easy test for that. Your body is reacting to something, that's clear, but it's not reacting via the IgE pathway. What you have is called localized allergic rhinitis. That's my diagnosis."

Basically, that means the immune cells lining the membranes of my nose and eyes react to allergens upon contact. The allergic reaction, in my case, is targeted or "localized" and not systemic or "generalized."

My skin cells and their antibodies might not be reacting to the pollen circulating in the spring air around me, but the cells that line my nose and eyes are. Unfortunately, this also means that there is no way to know which specific allergens are causing my symptoms. Strictly speaking, there is another method we could try, but it would involve placing a microscopic amount of each of the fifty allergens—one by one—directly onto the membranes of my eyes or nose and waiting for a physical reaction. Unsurprisingly, both Parikh and I were unwilling to do this.

Having exhausted all available methods, Parikh was unable to crack the case: The trigger of my allergies will remain a mystery. She wrote me prescriptions for a daily antihistamine nasal spray and eye drops. She advised me to stop taking oral antihistamines, since they have side effects and my allergies are localized. There's no need to risk the negative side effects of allergy medication circulating throughout my entire body if my allergies aren't a systemic problem. It's far better, she counseled, to target the treatment at the source of the symptoms.

By the end of this monthslong, very personal, and yet not so unusual tale of a complicated allergy diagnosis—replete with several negative allergy tests and based on a patient's history of self-reported and clinically observed symptoms—my question for you is this: Do I have a confirmed respiratory allergy or not?

The answer to this question depends on two things: The first is how we define what an allergy is and how we distinguish it from similar symptoms and medical conditions. Because I have low levels of IgE and no evidence of a systemic immune system reaction, but I *do* have activation of the immune cells lining my nose, eyes, and throat, then by the definition I gave you in chapter 1, I have an allergy, or Type I hypersensitivity, but I am *not* atopic. The second is the different types of evidence we'll accept as confirmation of a hyperactive immune reaction. If we were only going by the results of my clinical IgE skin and blood tests, then I would have no scientific "proof" of my allergy. If, however, we allow visible evidence of inflammation and irritation following exposure to pollen, then I do have evidence to corroborate my localized allergic response.

As my own story illustrates so well (perhaps too well), allergy diagnosis in the twenty-first century is a confusing maze. From the invention of skin scratch tests in 1865 to the recent development of fluorescent immunoassay tests for specific IgE antibodies, it's never been all that "easy" to diagnose or to medically confirm an allergic response without witnessing it firsthand. And the milder the reaction, or the more invisible, the more difficult it is to discover, diagnose, or "prove." In the remainder of this chapter, we'll take a look at the basic science behind deciphering our immune system and its reactions to common allergens. Diagnosing an allergy, as we'll see, relies just as much on honed skill and patient experience as it does on immunological science.

THE SHORT, LONG HISTORY OF ALLERGY TESTS

Allergy diagnosis has remained largely unchanged for well over a century. The methods and tests that an allergist uses today (highlighted by my own experience) would be familiar to any clinician working in 2000, 1970, 1930, and even—at least in the case of hay fever—as far back as 1865, the year that the British physician Dr. Charles Harrison Blackley invented the first skin-prick test. From the beginning of modern, organized allergy research (circa 1923, when the first professional association of allergists was formed), the standard diagnostic procedure has been to conduct (1) a thorough examination of the patient's medical history, including the onset of their symptoms, the timing of their allergy attacks, their occupation and home environment, and the frequency and duration of their symptoms; (2) a physical examination, recommended to rule out any other diseases that might have similar symptoms and to ascertain any complicating factors, such as other illnesses like diabetes, that might impact the patient's allergies; and (3) diagnostic tests, which have varied according to time period and available technology but have always included the ubiquitous skin-prick test.

By the 1930s, the allergist Dr. Warren T. Vaughan was arguing that every general physician should test his patients for allergies for their greater benefit. Vaughan knew that many patients who had persistent symptoms, or those whose maladies were not easily explained by other diagnoses, might be helped by specialized treatment and care. Vaughan advised complete honesty on the part of the patient when describing symptoms. Otherwise, he warned, patients were apt to receive not only the wrong diagnosis but also the wrong prescriptions for treatment and care of their conditions. Vaughan and other top allergists counseled against the hazards of self-diagnosis, urging those with symptoms to visit a trained specialist for testing.[3]

Vaughan's recommendations for respiratory and skin allergy testing in his 1931 book, simply titled *Allergy,* are thorough and represent the standard procedures and diagnostic tools of his time.[4] After taking the patient's history and conducting a physical exam, Vaughan would begin testing with a scratch test. Until the 1970s, when allergen extracts started to be mass-produced, allergists would craft their own for use in skin-prick tests and in immunotherapy treatment. Typically, the allergens used would represent the most common local pollens. If the scratch test failed, Vaughan recommended performing an intradermal test next. One might also use a subcutaneous test (an even deeper puncture that went below the skin layer) or an ophthalmic reaction (where a small amount of pollen powder would be placed inside the lower eyelid and then washed out after two or three minutes). If these tests produced inconclusive results, Vaughan counseled that the allergist should then conduct an intranasal test, where pollen would be blown into one side of the patient's nose to test for a reaction. (These last two are similar to what Parikh might have done had I chosen to ascertain what specific allergens triggered my own symptoms.) Next, the patient could be given a patch test, where pollen would be placed on the skin and then covered up for twelve to twenty-four hours. Vaughan counseled that the patch test was best for those with skin allergies, whose skin was often too sensitive and might react to the injections themselves. An allergist working in the 1930s and '40s might also perform a "passive transfer" test, using the P-K serum test discussed briefly in chapter 1. The nonallergic person would then be skin-tested for sensitivity; if they reacted, then the original patient was confirmed to have an allergy. The P-K test was most often done with infants and those who had severe skin eruptions that made it impossible to do the patch test.[5] If all else failed, Vaughan suggested that the allergist perform bacteriologic studies. After harvesting bacteria from all parts of the patient's body (teeth, sinus, intestines), the allergist would grow the bacterial sample and then use it as an extract to test for an allergic response in the patient. The same could be done for bronchial "secretions" or sputum, which could be collected, filtered, sterilized, and then used to try to inoculate the patient against an allergen. These tests were exhaustive and yet still might not produce any proof of an active allergy.

The private notebooks of William S. Thomas, an allergist working in New York City from the early 1920s to the late 1930s, are riddled with examples of what he called "skin test fallacy."[6] He noted that "Mrs. Keller is undoubtedly clinically sensitive to sheep wool and tobacco, but her skin reactions to these are negative." In the other direction: "Mr. Maresi gives a marked skin reaction to ragweed, but he has no hay fever or other allergic symptoms." And poor Mrs. Rushmore suffered terribly from ragweed and benefited from injections of ragweed pollen extract and yet always tested negative for ragweed in skin tests. In a book on allergy written in 1933, Dr. Samuel Feinberg counseled that one should never take the results of the standard skin tests as definitive of an allergy.[7] In his opinion, negative skin tests meant nothing; a patient might test negative and yet still be allergic (as indeed my own test results suggest is true). In 1931, Dr. Arthur Coca cautioned that many factors could affect the skin test: abnormal skin conditions, hot or cold temperatures, concentration of allergen, the time of the reaction, sensitive skin, the location on the body that the test was performed, the depth of the injection, and the proximity of the tests to one another.[8] Clearly, a lot could go wrong.

For food allergy, diagnostic tests were even more difficult to conduct. In the 1930s, most general physicians still thought food allergies were largely "imaginary."[9] In contrast, early researchers argued that food allergy was much more prevalent than assumed and posited that it might be the underlying cause of a variety of other less understood medical disorders.[10] In his 1931 book on food allergy, Dr. Albert Rowe argued that it was little understood and grossly underdiagnosed because it was harder to determine, since patients were largely nonreactive to food allergens in skin tests, and food allergies were generally milder than other allergies. (It's important to note that cases of anaphylaxis from the ingestion of food had yet to be officially recorded. They were suspected, but not yet proven, to exist. It's no longer accurate to suggest that food allergies are "milder" than other allergic conditions.) Rowe counseled that unlike "inhalant-type" allergies whose symptoms were fixed in the respiratory tract, food allergies could create symptoms anywhere in the body.[11] (This is somewhat accurate; food allergies can cause skin reactions and constrict airways.) That made relying on symptoms to diagnose a patient more difficult, since the symptoms of food allergy were similar to those of many other medical conditions.

Nevertheless, a diagnosis of food allergy could only be "proven" via the patient's own reports and the direct observation of negative reactions. Early food allergy patients were asked to go on strict elimination diets and carefully track their daily food intake to pinpoint the cause(s) of their allergy. Vaughan recommended that his patients keep detailed food diaries—a list of everything consumed in twenty-four hours. Once patients had experienced at least ten to twelve episodes of discomfort, they were instructed to bring their daily lists back to Vaughan for his analysis. Most patients would keep food diaries continuously for four weeks, recording all their symptoms, in addition to a "general diary" to record all events and emotions. Their allergists would use all this information to diagnose their food allergy—or to rule it out.

Though crude, these mid-twentieth-century diagnostic tools and tests went basically unchanged for decades. Still only moderately effective at diagnosing an allergy, the more modern version of the skin-prick test remains the standard.

TWENTY-FIRST-CENTURY DIAGNOSTICS:
DOING OUR BEST WITH WHAT WE HAVE

During interviews for this book, whenever I asked practitioners about the modern challenges of allergy diagnosis, and especially about the trickiness of using IgE tests as the marker of allergic disease, a lot of experts in the field told me that I should really be talking about all of this with Dr. Hugh Sampson. Sampson is the Kurt Hirschhorn Professor of Pediatrics at the Icahn School of Medicine at Mount Sinai and director emeritus of the Elliot and Roslyn Jaffe Food Allergy Institute in New York City. He was one of the first, and remains one of the most influential, people working seriously on food allergies in the United States. By the time he sat down to talk to me on the phone during the COVID-19 pandemic, Sampson had already been researching, diagnosing, and treating food allergy for forty years. In other words, he knows a thing or three.

I asked him about how things have changed over the last four decades. "Basically, when I started, allergists made the diagnosis of allergy by doing skin tests," he said. "The problem back then . . . well, and still today . . . is that you can have a positive skin test without getting clinical symptoms. So, at that time, when we were just looking at skin tests, it would turn out that for most foods, only about thirty to forty percent of people that had a positive skin test would actually react to that food."

Sampson reflected on the overall state of the field when he first began practicing and researching. In the early 1980s, allergy was still considered a backwater medical field. In fact, medical students barely got any training in allergy. (This is still the case; most doctors in training spend around two weeks studying allergic disease.) "People didn't really even think of it as a science," Sampson explained. "They didn't really believe the skin test meant anything."

There's something behind that lack of belief: It can often be a struggle to get accurate results from common skin-prick tests. First, skin tests have to be performed correctly, with both positive and negative controls. Negative controls are the diluents used in the mixtures and should result in zero response; positive controls are histamine, which normal skin will react to by forming a wheal. Second, skin tests and intradermal tests have to be administered precisely. For skin-prick tests for both respiratory and food allergies, the puncture from the applicator has to be deep enough to deliver the allergen far enough into the skin. But if it's too deep and the patient bleeds, that can be counted as a false positive (especially if the puncture is too deep in intradermal testing). If the scratches or injections are done too close together, the results can be hard to read, because it may be unclear which specific allergen caused a reaction. It's also far better if quality standardized allergen extracts are used, but this is more difficult than it might seem.

Part of the problem with the accuracy of skin tests is that several different companies currently manufacture the extracts used in skin-prick and intradermal tests—and those extracts can differ significantly in both their concentration of allergen (how much allergen is in each dose) and their makeup (what type of solution the allergen is mixed with). Because there are no regulations requiring standardization of commercial preparations of allergens for skin-prick tests, the quantities of allergen that are introduced can vary, so it's hard to know how much allergen has penetrated the skin. Not enough or too much can throw off the results. Sometimes, the inactive ingredients used in different extracts can cause a reaction themselves, leading to a false positive. Risk of injecting too much allergen is high for intradermal testing, which could lead to either false positives or more severe reactions. (In fact, all skin tests for allergies have to be performed in a clinical setting in case the patient has a severe reaction to one of the allergens.)

Recent studies on "the quality and potency of commercial extracts" in both the United States and Europe found particularly high variation in "extracts for mites, animal danders, molds, and pollens."[12] James Cook University in Australia found that the material used in testing for fish allergy is "unreliable."[13] The number of fish allergens contained in the solutions varied greatly, which might lead to false negatives. Currently, only four species of fish out of the hundreds of edible fish on our planet are tested. Allergen extracts used in most skin tests today are either one single allergen or a mixture of similar allergens (for example, a test for "grass" allergy will likely include multiple species of grass in the same extract). That can make it hard to accurately interpret results, especially if the extract is missing one of the prevalent types of vegetation found in the patient's geographic area. To make this process even more convoluted, the results of skin tests are collected, averaged, and then used to standardize the allergen extracts (this seems a bit like circular logic, but okay) that are then used to conduct epidemiological and pharmacological studies, which is one reason why we have trouble getting accurate numbers of allergy sufferers (more on that in chapter 3).

Even if everything is done correctly to produce high-quality allergen extracts, the reliability of skin-prick and intradermal test results can be affected by "the personnel's skills, test instruments, color of the skin, and potency of the extract," as well as "the site of testing, age, BMI, medications, allergen immunotherapy, circadian and seasonal variations, menstrual cycles, and stress and anxiety."[14] Taking antihistamines, steroids, antidepressants, tranquilizers, and other drugs that affect our immune system function can also throw off skin test results. Because of this, allergists typically ask patients to discontinue use of some of these drugs a few days to a week before testing. If the skin tests are conducted anyway, as would be the case for patients unable to interrupt their drug regimens for medical reasons, any negative test results have to be considered possible false negatives, even though positive results are still considered to be positive.

Infants are also notoriously hard to skin-test. Their skin doesn't show reactivity until they are around three months old, but even after that their results can be much harder to read and are considered more inconclusive than those of adults. This is why early-twentieth-century practitioners often defaulted to the use of P-K serum tests to detect allergen sensitivity for their infant patients.

Finally, and perhaps most important, there is currently no standardized or universally accepted system[15] for interpreting skin tests or for recording and collecting results. There are generalized suggestions available to practicing clinicians, but each allergist can determine for themselves how best to interpret the results of both skin-prick and intradermal allergen tests. This is why it is much better to have a trained allergist rather than a general practitioner administer and interpret skin tests. It can take years of experience to "read" tests with more accuracy.

Also, skin tests can only be performed on "normal," or currently nonreactive, skin; otherwise, they are nearly impossible to read. And, as you might imagine, this makes it very difficult for skin allergy patients to get accurate results at all.

When I spoke to Dr. Peter Lio, one of the top experts on atopic dermatitis (eczema), he explained that the common skin-prick tests are often not appropriate for patients with allergic skin conditions. Instead, in his practice, the skin tests are time intensive. Lio will place 80 to 120 stickers with a variety of allergens on a patient's back and leave them in place for forty-eight hours.

"It's kind of a drag," Lio said. "You put them on a patient on Monday. We take them off Wednesday. Then they come back in on Friday and we read the final ninety-six-hour reading. It's a bit more invasive for the patient, but it really does give us important information."

Once the final reading has been completed, and on the basis of any positive reactions, Lio then gives his patients a list of things to avoid in various products. Sometimes the triggers are hidden in shampoos, soaps, or other items used every day. It can take a while to know for certain which allergens are really causing the reactions because it can take up to two months after the patient stops coming into contact with those substances for their skin to calm down.

For patients to be diagnosed with atopic dermatitis in lieu of a positive skin test, three criteria must be met. First, the patient has to have an eczematous rash, or inflamed skin, not just blisters or bumps alone. Second, the patient has to have itch. Third, the rash and itch have to be chronic, or relapsing. One episode doesn't count. Atopic dermatitis is mostly diagnosed in children, and often clears as they age into adulthood, but can also become worse in adult patients.[16] Lio explained to me that current research may lead to the development of new diagnostic tests to subtype atopic dermatitis based on immunophenotyping (a test used to study the different proteins expressed by each cell). But for now, the patch test is really the only method he has for determining possible allergenic triggers for eczema.

For both respiratory and food allergies, tests for specific IgE antibody reactions to allergens are an option when skin test results are inconclusive or inconsistent. When Sampson first began his career, allergists were also using a radioallergosorbent test, or RAST, to check their patients' blood for IgE reactivity in response to different allergens. The test is a radioimmunoassay that uses a small amount of radioactive antigen mixed with a patient's blood serum. If the patient is allergic to the antigen, then the patient's IgE antibody will bind to the antigens; the free-floating antigens are measured by a gamma counter (the fewer the free-floating antigens, the more active the IgE, thus the more sensitive a patient would be to the antigen).

Today, the RAST has largely been replaced by newer immunoassay methods, but in common parlance the term "RAST" is still used—even by allergists—to refer to other blood tests. If, like me, you were to need a blood test, your allergist would typically order either an enzyme-linked immunosorbent assay (ELISA) or the more popular and more accurate fluorescence enzyme immunoassay (FEIA). With ELISA tests, an antigen and antibodies with attached enzyme markers are mixed with a patient's blood serum to detect antibody responses for specific allergens. ELISA tests are fast and very affordable but require the allergist to test individual allergens or groups of allergens separately. They also require a human to run them. FEIA tests use a similar methodology to both RAST and ELISA tests, only they use a fluorescent enzyme as an antibody marker to measure antibody response to specific antigens. FEIA tests are fully automated and less prone to errors and can screen against many allergens at once. The advantage of the standard FEIA test (the commercial name is ImmunoCAP) is that it can measure allergen-specific IgE (sIgE) instead of total blood serum IgE levels. It also lessens (but doesn't completely eliminate) the chance of getting a false positive due to the test accidentally picking up cross-reactivity, or your antibodies reacting to allergens that consist of proteins genetically similar to the ones you're actually allergic to (such as different nuts from the same family).

Yet even when a blood serum test "works" and shows a positive result for sIgE activity, it doesn't necessarily mean that a patient has an active allergy to that specific allergen, just that they show reactivity to that antigen. Sampson reminded me that relying on blood tests to diagnose food allergy is a very bad idea. He pointed out that when people who had positive blood test results are given an observed oral food challenge, "the rate of positive tests far exceeded the number of people who actually had clinical reactions." In fact, the rate of false positives of *both* skin and blood tests for food allergies can be as high as 50–60 percent.

As the decades passed, allergy researchers were eventually able to show strong correlations between the level of sIgE antibody in blood tests, the size of the wheal produced by a skin test, and the likelihood that a person would have an immune reaction upon ingesting a particular food or coming into contact with a respiratory or skin allergen. But this new understanding has also created some confusion among patients: They regularly conflate the level of IgE antibodies in their blood or the size of the wheal after a skin test with the severity of their allergy.[17] On social media sites like Reddit and Facebook, patients regularly share pictures of their skin tests to highlight how allergic they are. In other words, they equate tests that only measure a sensitivity or the probability of a reaction with being able to accurately assess the degree of allergic reaction that they will experience if they come into contact with the allergen in normal life. And, unfortunately, that's just not the case.

"There's not a good correlation between the size of [the wheal produced by] the skin test, or the level of antibody, and the severity of the reaction you'll have," Sampson explained to me. "The only thing those correlate with is the likelihood of reacting, not how bad that reaction might be."

That's why the gold standard for diagnosing a food allergy—both historically and today—is the double-blind placebo-controlled oral food challenge, often referred to as an OFC.

Despite the fact that OFCs are the best way of confirming a food allergy, they are the least likely tests to be performed. The reasons for this vary, but some of the most common are the cost of performing OFCs, because tests need to be performed in hospitals or other health-care locations that have the capacity to care for a patient experiencing anaphylaxis; the amount of time that it takes to complete OFCs, because each allergen needs to be tested separately and in increasing amounts over several days or weeks; and the very real risk that the tests may cause a severe reaction in patients, especially small children.[18] OFCs make parents particularly nervous and can cause a great deal of anxiety in children, too. In the absence of an OFC, most food allergies are diagnosed by using a combination of detailed medical history, physical exams, skin-prick tests, and sIgE blood tests. (Not recommended for use are intradermal tests because they can cause severe reactions; total serum IgE measurements, which gauge the presence of only generalized, not specific, allergic reactions; IgG measurements because everyone makes IgG in response to food proteins; or any other tests that might claim to assess food allergy.) Taken together, an experienced allergist can accurately diagnose most food allergies.[19] That said, without an OFC, there is no way to confirm with absolute certainty whether someone has a full-blown food allergy.

In addition to these challenges, Sampson pointed out that not enough testing has been done on adults. Most allergy research, especially in relationship to food, is conducted with young children (which makes sense, since most patients first develop food allergies in infancy or early childhood). This makes it harder to interpret study results for adults and can lead to confusion.

Food allergy diagnosis is further complicated by the fact that its major symptoms can mimic other gastrointestinal diseases or conditions that are not related to allergy at all. There are also food-related disorders that are not mediated by IgE at all, like food protein–induced enterocolitis syndrome, food protein–induced proctocolitis syndrome, and eosinophilic esophagitis (EoE).[20] Enterocolitis syndrome is an immune-induced inflammation of the intestines often triggered by cow's milk or grains that causes vomiting and diarrhea. Food protein–induced proctocolitis syndrome is an immune-induced inflammation of the colon often caused by cow's milk that can lead to bloody stools in infants. EoE is an inflammation caused by an excess of eosinophils (another type of white blood cell) lining the esophagus and triggered by specific foods. (We'll be taking a closer look at EoE in chapters 4 and 7.) These rare immune-mediated conditions (affecting approximately 0.5 percent, 0.12 percent, and 0.0005 percent of the general population respectively) often show up in infancy or early childhood but are not driven by the action of IgE antibodies. "And unfortunately," Sampson explained, "there is no good test for any of those."

Part of the problem with food allergy diagnosis and allergy diagnosis overall, Sampson told me, is that we still don't really understand the immunologic mechanisms behind many allergies. And, as allergy rates continue to rise, that means that we also don't have adequate diagnostic tools to meet the size of the problem.

A case in point is the skin-prick test. It remains the most ubiquitous, accessible, and cheapest test available for initial allergy diagnosis. But between 8 and 30 percent of the population[21] will have a positive skin test (or the development of a wheal) without exhibiting any symptoms of an allergy. Despite this, skin test results are still important indicators of allergy because studies have shown that between 30 and 60 percent of patients who show a sensitization[22] to a particular allergen will go on to develop an allergy. If you remember only one thing from this chapter, it should be this: Blood and skin tests only show sensitization to a particular allergen; they never confirm an allergy.[23] Any skin or respiratory allergy should be confirmed by an allergist using patient history and the presence of symptoms when the patient comes into contact with the allergen in the wild, so to speak.

The objective science of allergy diagnosis is peppered with subjectivity. Many allergists rely on their gut instincts, honed through years of clinical experience, to read skin test results and diagnose an allergy.

As Parikh suggested, interpreting allergy test results in the twenty-first century is an art form as much as it is a science.

GOOD TESTS, OLD TESTS, BAD TESTS, NEW TESTS

For the past few years, my good friend David has been experiencing generalized abdominal pain. About a year ago, he was diagnosed with a hernia and underwent two surgeries (the first one didn't work— a rarity, but it happens). David is typically a very healthy and happy person, but his extended illness, coupled with turning forty-five, put a dent in his otherwise optimistic armor. He doubled down on his yoga practice and committed to eating well. To that end, he went to see a naturopath. To figure out if David was allergic to anything he was eating, the naturopath wanted to order an IgG antibody blood test. An allergy, the naturopath surmised, might be contributing to his continued discomfort.

David was desperate to feel better, so he decided to take the blood test. Knowing that I was researching an entire book on allergies, he emailed me to ask me about his results. His IgG levels, he said, were high in response to several different foods. He was considering cutting all of them out of his diet but wanted my informed opinion first.

Over the years, I had heard from allergists that IgG antibody tests were, to put it frankly, total bunk. IgG makes up the lion's share of circulating antibodies in your bloodstream. It plays an important role in normal immune function and a part in some autoimmune disorders (as we saw in chapter 1) but not in Type I hypersensitivity, or allergy. Despite this, people like my friend David have been ordering commercially available IgG antibody tests in droves, hoping to find answers to a bevy of mysterious and unpleasant symptoms. Because the tests cannot actually tell the patients anything about their probability of having an allergy, most allergists find this new trend to be concerning. As Sampson explained, "The thing is that everybody makes IgG antibodies to food."

After we eat, as our stomachs begin to break down and digest our food, some of the natural proteins will pass through the gut barrier and enter our bloodstream. About 2 percent of the protein that we ingest on a daily basis will get into our blood circulation in what is called "immunogenic form." That simply means that it's capable of triggering the body's normal immune responses and activating our antibodies. Remember Dr. Avery August's description of our immune cells as curators of what things get to become part of us? IgG acts as that curator when it notices food proteins in our blood.

"So if you're eating eggs and drinking milk, you're going to have IgG antibodies to eggs and milk," Sampson said. "But there's never been a demonstration of pathogenesis [development of disease] related to those antibodies."

In other words, IgG doesn't cause food allergy or Type I allergic disease. Misunderstanding their IgG blood test results leads many people to eliminate basic—and nutritious—foods from their diet. Then, when they retake the blood test, their results show that their IgG levels have dropped. This is taken to be a sign that their efforts to avoid certain foods are paying off and that they are, indeed, "allergic" to those foods. In fact, there's zero evidence that IgG triggers any negative effects for the body. But it does make sense, at least from a physiological perspective, that if you avoid eating something, your body will stop producing IgG to it. That also means that you may be accidentally priming your antibodies to see those proteins as a problem if you eat them again in the future. There is growing evidence that IgG antibodies might actually be protective of an allergic reaction, since patients who go through immunotherapy for food allergy often generate significant increases of IgG antibodies in the process. As their bodies learn to tolerate small amounts of the proteins they are allergic to, their IgG levels go up. Sampson thinks that this is solid evidence that IgG antibodies likely play a role in normal, healthy immune function.

"If you're drinking milk and you don't have IgG antibodies to milk, then I start to worry about your immune system," Sampson said. He agrees with his peers that IgG tests are not only largely worthless in diagnosing allergies but should probably be banned for public use unless someone can prove their diagnostic validity. When I asked him why so many people still swear by the results, despite the complete lack of evidence that IgG causes allergy, he paused and then told me that he thinks there's a considerable placebo effect associated with the tests because they are so expensive to conduct. Depending on how many allergens are being tested, the cost can be hundreds of dollars. "When you spend that kind of money," Sampson mused, "you almost project that you feel better." If a person anticipates that they will feel worse after they ingest the foods they think are causing their symptoms, they actually do. It's the nocebo effect—the negative counterpart to the placebo effect.

When I wrote to my friend David that the consensus of all the allergy experts I've interviewed over the past several years is that IgG tests are at best useless or at worst dangerous, he wrote back to tell me that he trusts his naturopath and that he has felt better since he started avoiding gluten and dairy. While I kept trying to convince him, he kept sticking to his gut feeling (pun intended). I ended up feeling frustrated, but Sampson was sympathetic and not at all surprised by David's response. Sampson has seen it all before—many times.

"When I started in the area of food allergy," Sampson told me, "I spent all my time trying to convince people that a food *was* causing their symptoms. Now I spend all my time trying to convince people that a food *isn't* causing their symptoms. Everybody is doing all these tests, all these crazy tests. The problem is everybody eats about five or six times a day and you can always relate something back to a time when you ate. So, in this case, history can be very misleading."

Nonetheless, Sampson admitted that he didn't think oral allergy syndrome could possibly be real when he first heard about it decades ago. Not as serious as a full-blown food allergy, oral allergy syndrome is related to seasonal pollen allergies. Whenever a person with oral allergy eats certain fruits or vegetables, their immune system recognizes the produce's molecular structure as being similar to a type of pollen they are allergic to, and their mouth tingles or itches. It seemed like an improbable reaction to Sampson but turned out to be a real phenomenon. So while it's extremely unlikely, it's not entirely impossible that IgG may play a small role in some of the allergic disorders. More than once, Sampson has been perfectly happy to prove himself wrong in his own research, and he told me that there is much that we still don't understand about allergic immune responses. There could be conditions and triggers that we don't know about yet (as we'll discover in chapter 6, when we discuss the relatively new "meat allergy"). But Sampson finds the increase in attention and funding that allergy has received over the past twenty years encouraging. He is hopeful that ongoing research will eventually eradicate all allergies—but he doesn't see that happening anytime soon. Not in his lifetime and maybe not in mine. All we can do for now, he said, is find ways to turn down the immune response, not stop it altogether. And to do that, we need to keep working on developing better diagnostic tools to detect allergy.

IRRITATED PATIENTS AND PHYSICIANS

If by now you're not feeling particularly confident about the current diagnostic tools used to adjudicate your allergy, you're not alone. Allergists are often equally frustrated by the tools available to them and are hopeful about the development of better, more precise methods to test for allergic responses. Dr. Ruchi Gupta, a pediatrician and an epidemiologist at Northwestern University, pointed out to me that our current diagnostic tests for allergies are very good at predicting if you don't have an allergy, but very poor at predicting whether you do.

"The negative predictive value is very high, but the positive predictive value is very low," Gupta said. "It's almost like a flip of a coin for positive predictive value. So, if you have a positive test, it's almost like you have a fifty percent chance of having that food allergy and a fifty percent chance of not having it."

That's not very comforting, I know.

Ideally, future tests would include methods that do not rely as heavily upon IgE as a sign of a true allergic response, especially since

IgE reactions don't always indicate the presence of an allergy and there are many non-IgE-mediated allergies that cannot be tested via current methods.

Part of the problem with diagnostic tests for allergy is that they are related to two fundamental problems with scientific research itself. The first is that there are limits to scientific technology—to what we can see and what we can study. The second is that all scientific knowledge relies on an understanding of averages.

"When we're testing blood," Dr. Alkis Togias at the NIH explained, "there are billions of cells in the blood and many times we're looking at an average response of those cells to something, or an average expression of a particular molecule. That average leaves behind a lot to learn. There could be, for example, a number of cells within the same individual that are very low. We look at the average and of course we miss that there are two populations of cells."

To put that another way, some cells might be responding to a particular allergen, but some aren't. The blood test result is the average reaction of all those cells but hides the fact that some cells might be very reactive while others are not reactive at all. What this really means is that your blood test might show negative results even if *some* of your cells responded positively to the allergen, or vice versa.

On the bright side, Togias also reminded me that researchers at the NIH and around the world are hard at work developing new molecular tools to help diagnose allergy. Even so, they will likely be more expensive and their use will be limited, especially for those without adequate access to healthcare or the resources to pay for them out of pocket. For the foreseeable future, we remain reliant on the tests described above for the majority of our allergy diagnoses.

But if, as we saw in chapter 1, allergy itself can be a fuzzy term, and allergy diagnosis is complicated at best, then how do we assess just how large the global problem of allergy really is?

CHAPTER 3

Our Allergic World:
Measuring the Rise of Allergic Disease

———

THE PROBLEM OF UNCERTAIN NUMBERS

Allergies are never quite what they seem. They are hard to pin down. Often harder to diagnose. And even harder to measure.

Accurate measurements of the incidence rates of allergic conditions matter. Numbers drive everything in medical research, from the allocation of funds to the development of new pharmaceuticals. To begin to grasp just how big the problem might be, and why allergies may be the defining chronic medical condition of the twenty-first century, we have to dive headfirst into a sea of statistics. The following figures, culled from some of the latest available data, highlight how pervasive and widespread allergies are today. As of this writing:

- An estimated 235 million people worldwide suffer from asthma.
- Globally, 240 to 550 million people may suffer from a food allergy.
- Drug allergy may affect up to 10 percent of the world's population and up to 20 percent of all hospitalized patients worldwide.
- Between 10 and 30 percent of the world's population has allergic rhinitis (hay fever).
- At least one form of allergic disease afflicts 20–30 percent of the total population of India.

- Respiratory allergy affects 33 percent of Indians.
- Some form of chronic allergy challenges 150 million Europeans.
- Half of Ugandans have an allergy.
- Food allergy impacts 7.7 percent of Chinese children.

These numbers are, on some level, incomprehensible in their scope, and yet we see figures like them every day. Most of us have become so accustomed to seeing tables, graphs, survey results, and percentages blazoned across our news feeds, that facts and numbers manage to intrigue us, overwhelm us, numb us, and bore us all at the same time. We live in the era of big data, global science, and Excel spreadsheets, where, as Joseph Stalin is purported to have said, "If only one man dies of hunger, that is a tragedy. If millions die, that's only statistics." If we translate this logic into the realm of modern-day medical conditions, then perhaps we can start to understand more clearly why we haven't paid enough attention to these staggering rates: If only one child dies of anaphylaxis from eating a peanut or from an attack of severe allergic asthma, that is a tragedy. But if millions more suffer from a food allergy or asthma and don't die, that's just statistics. And while massive numbers like these might be able to tell us a lot about the sheer scope of the global problem of allergy, they can't tell us everything that we need to know.

We have a difficult time visualizing all the allergy sufferers, and their collective day-to-day struggles with their conditions, that constitute all this data in the first place. Individual stories—like my father's, mine, and perhaps your own—tend to get lost. Important details and contextual information—all the lived experiences of the billions of people who have allergies—drop out of the data sets.

Take Veronica. Veronica is a vivacious woman in her early thirties whose respiratory allergies are so severe that she dreads the first hints of spring. Warming temperatures, green shoots thrusting out of the ground, gradually longer days, budding trees—all these spell disaster for Veronica if she hasn't started to take her prescribed allergy medication early enough. Thanks to the vagaries of a changing climate, each year feels more and more like a guessing game: When will spring arrive? Veronica tries to make an appointment with her regular doctor three or four weeks before spring begins in earnest. And yet even when she times everything perfectly, her allergies can still be unpredictable. If it's a particularly bad pollen year—if pollen loads are markedly higher or if the pollen season is longer than usual—then Veronica still suffers, even on her prescription antihistamines.

"When I walk to work, I have to make sure I have my wraparound sunglasses on," she explained to me one afternoon, as we were sitting comfortably in her office. "My eyes are my trigger points. If I forget to wear them, then I look like I've been crying, or I've been up all night partying. Either way, it's not a good look at work."

Veronica showers after she gets home each day to wash the pollen out of her hair, avoids outdoor events on days when the pollen count is supposed to be particularly high, and generally feels fatigued for three or four months out of each year. When I asked if her husband, friends, and family are understanding, she nodded and said, "My whole family has allergies, so they get it. Everyone's on some form of Claritin or Allegra or Xyzal, all across the board." And lately, she said, everyone's allergies seem to have been getting worse. As long as her allergy medication keeps working, it's fine. But she worries about what will happen to her when even the best available prescription medications aren't effective anymore.

———

When I first started to look through the statistics, I was both overwhelmed and confused. Exactly what were official numbers based on, and why did they often vary or reflect such a wide range of possibilities? Obviously, all statistics are estimates. They are calculated based on smaller, representative sample sizes. But I wanted more details about who was doing the sampling and how, so I reached out to the U.S. Centers for Disease Control and Prevention (CDC) to try to find some answers. The CDC keeps track of both asthma and food allergy rates because those are the two deadliest forms of allergic disease and are likeliest to contribute to our national rates of mortality. Yet after several rounds of phone calls and emails with staff members, I had no answers. After doing a little more legwork and a lot more interviews with allergy researchers, I realized that's because it's difficult—if not impossible—to know with any absolute certainty just how many people are afflicted by allergy. It's equally hard to definitively answer the question that everyone wants to know the answer to: Are things getting worse?

This was one of the burning questions I had after I was diagnosed myself and began talking to others about their own allergies. Allergy experts, healthcare providers, pharmaceutical and biotech companies, nonallergic citizens, allergy sufferers like Veronica and quite possibly you, concerned reader, all want to know if allergies are more prevalent now than they were in the past and if the rates of allergies will continue to rise in the foreseeable future. Are all the numbers truly worse than they were ten years ago or twenty or thirty? Are the rates of allergy really rising decade after decade, or have new public health awareness campaigns and more accurate diagnostic tools simply made us better at spotting and diagnosing allergies, thereby bumping up the numbers? Are people living in the twenty-first century actually more likely to develop allergies or experience them more frequently and with more severe symptoms?

I spent over five years researching and writing this book, reading about allergies in our past, interviewing allergists, and visiting scientific labs working on allergies. I asked everyone I met if they think allergies are both becoming more prevalent in the general population and becoming more severe in nature. Nearly everyone said yes to both questions; yet, they also cautioned that we're just at the beginning of our journey to understand allergies from a scientific vantage point and that the data we currently have isn't as good as it could, or should, be.

Allergy experts who have been working in the field for decades all told me the same thing: It's difficult to accurately assess the current situation because it's difficult to get reliable data about the number of allergy sufferers. On the one hand, we have countless individual narratives of people suffering from different forms of allergic disease—eczema, asthma, hay fever, food allergy—and clinical notes and diagnoses from physicians or allergists. On the other hand, we have the compiled and tabulated official statistics. When you dig into those epidemiological studies, you quickly see a few glaring problems.

To start with, the definition of what an allergy is—or, perhaps more critically, what it isn't—can affect how people are counted, throwing off the accuracy of the statistics. Disease categories are not stable entities or "things" in the world; they are descriptions of a collection of typical symptoms and biological signs of disease. Even something seemingly "easy" to define—like asthma—is more convoluted than at first glance. The official definition of asthma has shifted multiple times since the 1950s. Epidemiological studies don't always use the same markers of disease, so someone who counts as asthmatic in one study may not qualify as asthmatic in another. In one metastudy, researchers found that 122 studies of asthma prevalence in children did not use standardized definitions or symptoms of asthma, making the data impossible to compile or compare.[1] In fact, sixty different definitions of asthma were used across the 122 studies. When the four most popular definitions were applied to the same set of data, the variation in how many children could be categorized as "asthmatic" was astounding. Depending on which definition was used, up to 39 percent of children went from having asthma to not having it.

So do the children in these studies have asthma or not? And who decides? The parents who witness their children slightly wheezing at the playground or having difficulty breathing at bedtime? The pediatricians who take a family history and then use a spirometer to measure their young patients' lung function? Or the epidemiologists looking at insurance claims coding for asthma, the number of inhaler prescriptions being written, or self-reported survey data of parents with children under the age of eighteen? This is what makes epidemiological data about how many people have allergy so difficult to collect, to decipher, and to write about.

Dr. Neeru Khurana Hershey, a physician and an asthma researcher at Cincinnati Children's Hospital with decades of experience, explained to me why allergic asthma, in particular, is so hard to track. "Asthma is a garbage term," she said. "It's the name of a symptom, not a disease. Asthma is heterogeneous. It is defined by a constellation of symptoms, which can result from different paths." In other words, many different medical conditions can cause an asthmatic reaction, not just allergy. That, Khurana Hershey explained, makes it hard to measure allergic asthma specifically—or to separate allergies out from asthma's other causes like exercise or other pulmonary conditions. To complicate matters further, even if allergies aren't the root cause of a patient's asthma, they may still be an environmental trigger for an asthma attack. Unless you look at each patient's medical history, it's impossible to tell who has "allergic" asthma and who has "nonallergic" asthma with allergic triggers.

And this isn't only a problem for asthma.

The definitions of nearly all the different forms of allergic conditions used to compile official data on global rates of allergy are fuzzy, contested, and constantly shifting. Surprisingly, hay fever—the oldest medically recognized allergy—is much more difficult to define than one might first assume, and the symptoms used to measure it can vary widely. And even if studies are rigorous and rely on clinical tests or an official diagnosis in order to be counted as a confirmed diagnosis (and most don't), the resulting numbers still depend on how the researchers initially defined the disease categories in the first place. All of this is, to put it mildly, confusing and frustrating, and it often leads to a big discrepancy in the official numbers of allergy sufferers.

Here's a pertinent example of how difficult it is to get more precise numbers about how many people are sniffling, sneezing, and otherwise irritated. Measurements of allergic rhinitis range from 10 to 40 percent of the total world population. On a global scale, the difference between 10 and 40 percent is enormous—it's like adding or subtracting the population of an entire continent. The large variance here is due to differences in definitions of what constitutes hay fever, the diagnostic criteria used to assess the condition in individual and national surveys (such as watery eyes or frequent sneezing), and the subject populations being measured (which socioeconomic groups and geographic areas are represented in the survey data being compiled).

To begin with, not everyone who has hay fever gets tested for it, and people who self-diagnose aren't always reflected in official figures. Even when people suffering from allergic rhinitis go to see their doctor or a general practitioner, they might not leave with the correct diagnosis. Additionally, not everyone who has an allergy knows that they have one or would identify as an allergy sufferer, especially if symptoms are mild or exposure to the allergen is rare. My dad didn't know about his bee venom allergy and I didn't know about my respiratory allergies, and neither one of us would have ticked the box next to "allergy" on our family medical history or answered yes to a survey question about having an allergy. And that's often how we collect data on the rate of allergies in the first place—we ask people directly or survey them about their symptoms.

This is a major issue with the reliability and accuracy of our current allergy numbers. Most epidemiological studies on allergies are based on self-reporting of symptoms via Web-based or telephone surveys. We rely on the people who have allergies to accurately assess their own symptoms and to truthfully report them so that their responses can be sorted into the right category and counted. The glaring problem with this approach is that the symptoms of allergy are often similar or identical to those of other medical conditions and therefore can be confusing. Self-reported symptoms are, at best, evidence that a patient *may have* an underlying allergy. Without a medical diagnosis, self-reported symptoms alone cannot be used as confirmation of a true allergic response.

WHAT THE DATA CAN TELL US—AND WHAT IT CAN'T

Even if we err on the side of caution for our estimates and argue that only 10 percent of the global population will have a respiratory allergy at any time in their lives, that is still a nearly unfathomable number—currently 800 million people worldwide.

So what do we know about this figure and the people, like Victoria, it represents? Unlike a food allergy, which is sometimes outgrown, respiratory allergies are usually chronic for life. That means that those numbers probably won't change throughout a generation. We also know that most patients with respiratory allergies have bad enough symptoms that they regularly use over-the-counter (three out of every four people) or prescription (half of those afflicted) medications.[2]

Allergic sinusitis (the inflammation of the nasal cavity caused by an underlying respiratory allergy) costs Americans around $6 billion per year in healthcare expenditures.[3] Americans also miss approximately 3.8 million days of work and school each year due to their respiratory allergies.[4] Patients who have moderate to severe respiratory allergies report significant reductions in their quality of life, including "disrupted sleep pattern, fatigue, and poor concentration."[5] In fact, 59 percent of allergy sufferers in a recent survey said that their nasal congestion had negatively impacted their ability to concentrate at work, leading to poor productivity, and about 80 percent of allergy sufferers have trouble sleeping at night, leading to increased fatigue during the day.[6] The physical symptoms of allergies lead to emotional effects like frustration.

Interestingly, Gallup reports that more Americans report being sick with allergies during the winter months than with either a cold or the flu; about 10 percent of the general population suffers from winter allergies. Gallup data also suggests that women are far more likely to report their allergies than men—which could be due to the stigma associated with allergies.[7] We often view those with allergies as somehow "weaker" than their nonallergic counterparts. Both the highest and lowest income brackets report more allergies than those in the middle, and people living in the South report more allergies than any other region in the United States.

In sum, the available data on allergies can tell us a lot, but it can't tell us everything, and it doesn't always tell us what we most want—or need—to know. Accuracy in this data matters for a whole host of reasons, so it's important that we find ways to get more precise statistics. Accurate numbers lead to better decision-making about which allergies to focus research funds on—right now, that's asthma and food allergy, leaving hay fever, atopic dermatitis, and contact, drug, insect, and occupational allergies in the dust (so to speak). With limited resources available, it's typically the things that kill us that epidemiologists and other public health officials are most interested in tracking. People with severe hay fever, like Veronica, might vociferously disagree with that assessment, since they know firsthand that while the condition might not kill someone, it can seriously interfere with a patient's quality of life. Scientific research dollars often lead to the discovery of biological mechanisms that can be turned into better treatments for patients.

THE DATA DETECTIVE

No one knows how important having more accurate numbers is better than Dr. Ruchi Gupta. Gupta is the director of the Center for Food Allergy & Asthma Research at Northwestern's Institute for Public Health and Medicine. She's also a practicing pediatrician at the Ann & Robert H. Lurie Children's Hospital of Chicago, with more than sixteen years' worth of experience studying and treating allergies. And she's a mom whose child has a severe food allergy, so she has a highly personal stake in her research.

Gupta began her medical career studying asthma and then got interested in food allergies. After receiving her master's degree in public health, she was lured to Chicago by the opportunity to study with one of the world leaders in asthma research. At first, she focused her own research on disparities in asthma care. Then she met a family that was dealing with a bevy of allergic diseases, including asthma and food allergy. The family complained about the lack of information on food allergy and Gupta became intrigued. Right away, she noticed that people working in food allergies didn't have great data available to them.

"Compared to asthma research, we knew so little about food allergies," Gupta explained. "To the point that there were no prevalence numbers being collected in the United States, so it was very unclear how many people were being impacted."

Gupta argues that we're mostly measuring those allergy sufferers who can afford to go see a doctor, or who live in urban areas, or who have good healthcare coverage. People without adequate access to healthcare in the first place, or those living in more rural areas, might not be counted in these official statistics at all. All of these factors can throw off official figures. And if you use surveys, or ask people to self-report their symptoms, then some of the people reporting as having allergies may not, in fact, have them. Overestimation and underestimation are chronic problems—especially for food allergies.

Increased press coverage and recent media attention on the topic of food allergies has only led to increased confusion among the general population. Awareness campaigns have done their job too well. Today, if people experience a symptom like lower abdominal pain soon after eating, they often attribute it to an underlying food allergy—when it could very well be something else entirely. Many different medical conditions have similar symptoms.

"There's intolerance, oral allergy syndrome, celiac, and Crohn's disease. There are so many GI conditions that you could have, but if you ingest the food and you have a negative reaction, it's hard to know if that was food allergy or food poisoning or just an intolerance," Gupta said. "It's hard for people to tell what is happening in their body."

Gupta partially blames the fuzziness of the term "allergy" itself. It's imprecise and it covers a wide spectrum—from mild sniffles to anaphylaxis—so it's a very confusing term for most people.

To try to correct for these gaps in the available data, Gupta and her research team devised a comprehensive survey that drills down into the nitty-gritty details, asking exhaustive questions about people's symptoms and daily experiences. Patients' answers make it easier to throw out any responses that don't clearly indicate a possible food allergy. It's a conservative method, but one that makes Gupta more confident about the data she's collected, though she admits that even her numbers could be wrong. There's just no way to say for sure, without doing oral food challenges, the gold standard for testing to confirm food allergies, but she insists that the numbers she's seeing from her surveys are still incredibly significant.

What Gupta knows for certain, she tells me, is that the problem of allergies is already considerable and seems to be getting worse with each passing decade. Her statistics are concerning, and surprising. According to Gupta's latest survey results (released in 2019), up to 10.8 percent of Americans show convincing evidence of a food allergy.[8] Nearly double that number, or 19 percent, self-identified as having an allergy, but only 5 percent of all respondents had a physician-confirmed diagnosis for food allergy. Using data compiled from recent studies, other well-regarded researchers have estimated that "food allergy likely affects nearly 5 percent of adults and 8 percent of children, with growing evidence of an increase in prevalence."[9]

After walking me through all the various problems with data collection, Gupta asked, "So which number do you believe?"

Ultimately, Gupta hopes that future collection of clinical data on a massive scale—or the rise of big data in healthcare—could help to solve the puzzle of allergies and give clinicians a better picture of the scope of the problem. But, at least for now, we're stuck with largely unreliable data—and a lot of questions about the true scope of an already massive epidemic of allergic disease.

A WORSENING EPIDEMIC

Though researchers may disagree about definitions and symptoms and methodology, they can all agree on one thing: Allergies have gotten worse over the past few decades, and the numbers of allergy sufferers worldwide is likely to keep growing apace. Looking at the data that we do have from the last century, there's a consensus that hay fever rates in the United States increased in the mid-twentieth century.[10] Data suggests that the incidence of asthma increased beginning in the 1960s and peaked sometime in the 1990s. Since then, rates of asthma have remained fairly constant. For respiratory allergic diseases and atopic sensitization (skin allergy), the levels likely increased over the past few decades as geographic differences in prevalence rates shrank. For instance, rates of atopic disease doubled in Ghana between 1993 and 2003.[11] For food allergies, the rise in global incidence rates has been the most dramatic and visible, beginning in earnest in the 1990s and growing steadily ever since.

Dr. Scott Sicherer, the director of the Elliot and Roslyn Jaffe Food Allergy Institute and the Elliot and Roslyn Jaffe Professor of Pediatric Allergy at Mount Sinai Hospital in New York City, has seen the rise of food allergy up close. When he began work at the Jaffe Institute in 1997, his team conducted a study in collaboration with the Food Allergy and Anaphylaxis Network that showed one in 250 children were reporting a food allergy to peanuts or tree nuts. By 2008, Sicherer's work had shown that rate had more than tripled to one in 70.

"I didn't believe the 2008 study at first," he told me. Sicherer initially thought the rate reflected a problem with the study's methodology— that is, until he saw similar numbers coming in from Canada, Australia, and the United Kingdom, all showing that about 1 percent or more of children had peanut allergy. Today, Sicherer has no doubt that allergies have increased over the past few decades.

"We're also seeing less food allergy being outgrown and more emerging," Sicherer said. "The severity might be intrinsically no different than it was twenty years ago, but with more people affected, it's a big deal."[12]

While all of this data is compelling, perhaps the most convincing evidence we have for the rise of allergies over the past thirty years is hospital admissions. Every two hours, someone with a severe allergy ends up in the emergency room. Those numbers seem like incontrovertible evidence that the problem of allergic diseases is expanding.

According to researchers at Imperial College London, who scoured available data over the past two decades, food anaphylaxis hospital admissions increased 5.7 percent (from 1998 to 2018), while fatalities decreased from 0.70 to 0.19 percent.[13] During the same period, prescriptions for adrenaline auto-injectors, or EpiPens, increased by 336 percent. The researchers controlled for changes to the definitions and criteria for food anaphylaxis and think that improved diagnosis and management of food allergy led to the decrease in deaths even as overall incident rates rose.

Hospital admissions for asthma tripled in just two decades, between the 1970s and the 1990s, before leveling off to a steady rate today.[14] And although rates of asthma in developed nations have been slowing, they continue to climb in underdeveloped parts of the world, causing the overall rate to continue its rise globally, even while it remains constant in places like the United States.

This is why experts predict that allergy rates will continue to rise for the next several decades. Allergic disease is less prevalent in rural areas of low-income countries, but allergic sensitization is at the same level (quick reminder: you can have a sensitivity without developing an allergy). In other words, people everywhere have the same sensitization, but there are fewer active symptoms and fewer cases of active disease in rural areas of poor countries. As countries begin to develop, rates of allergy tend to rise. Why?

———

As I sit down to write this conclusion, it is summer in Brooklyn, where I live close to a large, gorgeous city park. I go on long walks there almost every day if it is not too rainy or too sweltering or the air is not too clogged with pollution. Some days, I have no symptoms related to my localized allergic rhinitis at all and I enjoy the park with impunity. Other days are almost unbearable. When I return home, my eyes sting and itch and if I dare try to touch or gently rub them, I trigger a sneezing fit that may last as long as thirty minutes. Sometimes the surface of my eyeballs burns so much that my eyelids reflexively pinch shut, my conjunctivas producing so many involuntary tears that it looks as if I'm in the middle of a nasty emotional crying jag.

On the bad days, I open a weather app on my phone and check pollen counts, doing some quasi-scientific sleuthing to see what might be causing my intermittent misery. It always says the same thing: very high levels of grass. I assume I must be allergic to one of the species of grass in my area, but who knows which one.

At the time I'm writing this, we are also knee-deep in a global pandemic. COVID-19 is making most allergies look inconsequential by comparison. Every time I sneeze more than once or my throat feels a little scratchy, both extremely normal symptoms of seasonal respiratory allergies, I feel a small mote of panic start to grip me. Is it allergies? Or a sign that I have the more dreaded SARS-CoV-2? Normal allergy symptoms don't seem at all "normal" right now. But then again, they never really were normal. Allergy symptoms have always been a sign of something gone amiss.

Our collective allergy symptoms—all our runny noses, scratchy eyes, raw skin, upset stomachs, unhappy intestines, swollen esophagi, irritated lungs, and difficulty breathing—are trying to tell us something important about the overall health of our immune systems in the twenty-first century, about how we live our lives, about how overwhelmed our cells often are by our environment. And while it's true that the scientific definition of allergy has shifted more than once since its inception one hundred years ago, as we've seen, what hasn't changed is the dip in the quality of life that all those symptoms have caused for millions of allergy sufferers worldwide. As our understanding of immune function evolves, so, too, do the ways in which we talk about, categorize, and treat allergic diseases. We know more about allergy and our immune systems than we ever have, yet there is much about our basic immune system function that still eludes us. We use the same basic diagnostic tools for allergy—more or less—that we have for well over a century now, and we do the best with them that we can. Scientists the world over are working very hard, even as I write this during a pandemic, to better understand how our cells learn to tolerate the billions of visible and invisible things around us every day. Each advance in that knowledge will shift the boundaries of what we think of as an allergy and may give birth to new allergic disorders we haven't even thought about yet. All that new knowledge will also help biomedical engineers to invent new diagnostic tests or innovate older versions to deliver more precise and accurate results. At the very least, one can hope that the future of allergy medicine will look very different from its past and its present. (We'll look more deeply into these possible futures near the end of the book.)

Despite all the confusion and messiness surrounding the definition of allergies and their diagnosis, we know one thing with absolute certainty: Allergies—whatever we call them and however we define them—have gotten consistently worse over the past two hundred years and show no signs of abating. We also know that people are experiencing worse symptoms and longer allergy seasons. We are in the midst of a growing global epidemic of allergic disease. The next section of this book will attempt to answer a single, overarching question.

Why?

PART TWO

Theories

○ ○ ○

Despite the confusion and messiness of sorting out who does and doesn't have an allergy, it's clear to most allergy experts and public health epidemiologists that the overall rate of allergy continues to rise. In fact, the documented cases of all forms of allergy have been steadily increasing since the start of the Industrial Revolution in the early nineteenth century. If allergies have been on the rise for the past two centuries and show no signs of abating, then the next logical question to ask is: *Why?* Part 2 explores some of the most popular scientific (and a few nonscientific) theories that attempt to explain the modern allergy epidemic.

CHAPTER 4

Allergic Inheritance: Allergies as a "Normal" Immune Response

———

It's the middle of the COVID-19 pandemic and traveling to meet allergy researchers is impossible. As I go through my notes and questions, I'm waiting for Dr. Somnath Mukhopadhyay, chair of Pediatrics at Brighton and Sussex Medical School in the United Kingdom, to enter the virtual meeting room. Mukhopadhyay has been researching allergies for twenty years; specifically, he looks for possible genes or gene segments that might be linked to the development of allergies in young children. We are scheduled to discuss some of his most recent studies showing a correlation between a genetic skin barrier defect and a higher risk for developing allergies. This finding could help explain what is known as the atopic march—the sequential development in young children of atopic dermatitis, often followed by food allergy and/or asthma. This progression from skin allergy to food and/or respiratory allergy and asthma has been well documented in young children, but its ultimate cause remains a mystery. Most researchers expect that individual genetics plays a role in allergies, and Mukhopadhyay was one of the first to look for biological clues, in a large data set of genetic information on young patients in Scotland.

When he enters the virtual room, Mukhopadhyay is already smiling. It's clear after the first few minutes that he is passionate about his work and eager to talk shop. After I quickly explain my own research and experiences with allergies, he politely interrupts me, leaning forward into his camera.

"I see," he says, as his face becomes more serious. "So you are really here today because you want to know why your father died."

I hadn't really thought of my interest in the genetic causes of allergy in relationship to my own family's history. I had been interested in figuring out the primary cause of all our irritations and inflammation, and human biology—specifically our genetics—seemed an obvious place to start. My logic had been simple: Maybe something inside us is ultimately responsible for all of our allergies and, if so, then I wanted to ferret it out. But I quickly realized that of course Mukhopadhyay was right, at least to a certain extent. I did want to find out if my father's genes were ultimately responsible for his death and a part of my own family inheritance.

"Yes," I respond after a moment's pause, "I suppose I am."

Mukhopadhyay nods and looks straight at his lens. I can almost feel his compassion through the screen. He is intently concentrated on me, as if I were one of his patients.

"Bees are stinging millions of people each year, Theresa," Mukhopadhyay says. "And you have to think, Why did my dad die? And that question has not been answered."

It's clear he wants to answer this question for me as carefully and thoroughly as he can. He knows the answer is important not only to me because of my father's untimely death but also to everyone else around the world who might be coping with severe allergies. The questions of why we have allergies in the first place, why some of us develop them and some of us do not, and why some of us suffer so greatly are perhaps the central questions of this entire book. In fact, they may be the central medical questions of this century. Despite more than a hundred years of immunological science, we humans still don't really understand our immune systems, and as our world changes, comprehending how our bodies respond to altered environments may be critical to our survival. If COVID-19 has taught us anything, it's that immune function is the difference between living well, suffering, and dying miserably.

In my father's case, we might simplistically say he died because he was sitting upright in his car, or because he didn't have an EpiPen, or because he didn't get care quickly enough. But that, Mukhopadhyay suggests, is not really what I want to know. What I—and others like me—want to know is why *my* dad, why *that* moment, why *his* biological response, when so many people are stung each year and easily survive. In the wake of a tragedy, one craves an explanation that makes sense out of otherwise seemingly random events. We want to reduce the complexity of an individual's death to a simple biological answer because a biological problem might be fixable—or at least preventable.

"The answer," Mukhopadhyay says, "lies in the fact that your father's body reacted to the bee sting in a completely different manner to the way millions of people are experiencing and managing and tackling bee stings inside their bodies. And this very important question of *why* is key to the management of allergies. You might have allergies; I might have allergies. But we may have completely different biological reasons and responses."

As we saw in the last few chapters, allergies can be tricky. Symptoms are changeable and no two allergy patients are exactly alike. That stands to reason, since each individual's immune system cells are responsible for making decisions about how to react to the variety of organisms, chemicals, and proteins that they come into contact with on a daily basis. Each person's cells will respond differently to the same stimulus. Dr. Avery August, professor of immunology at Cornell University, told me that sometimes even identical cells within the same person will respond differently to the exact same exposure. The genetics is the same, the exposure is the same, and the lifestyle is the same—yet one T cell might decide to overreact to a peanut protein while another T cell completely ignores the same protein after initial contact. And no one, August said, knows why any particular individual immune cell makes the decision it makes. If enough of your body's cells decide that an otherwise innocuous substance in or on your body is dangerous, you'll experience an allergic reaction. But even in a person having a moderate to severe allergy attack, some of their immune cells will choose *not* to react, all but ignoring the triggering substance. Like everything else to do with understanding our immune responses, the biological causes of allergies are mired in mystery and hard to tease out from other confounding causes.

We are about to embark on a historical journey from the beginnings of basic allergology to modern scientific research into the biological underpinnings of why our immune systems developed an ability to accidentally kill us in the first place. Evolution is conservative. It tends to save DNA that gives human beings—and every other species, for that matter—a greater chance at survival. Then how might we explain our immune system's capacity to overreact to basic food groups or plant pollen? Why would a biological system designed to protect us from harmful bacteria, viruses, and parasites also be able to cause so much havoc in response to something as innocuous as a dust mite or speck of cat dander? The answer lies in the complex web of interactions between our genes, our inherited genetic variations, our immune cells, and our environments.

THE DISCOVERY OF THE DARK SIDE OF IMMUNE FUNCTION

In the late 1800s and early 1900s, the concept of immunity was all the rage. The success of the new germ theory of disease had led to the fruitful development of vaccines against many common infectious diseases, including smallpox, cholera, and rabies. Scientists already knew that immunity was, at its base, about triggering the body's natural defenses, but they hadn't yet discovered the darker side of immune function. By the turn of the twentieth century, it wouldn't have been all that strange for scientists working in the nascent field of immunology to expect that they might be able to create immunity to a wide variety of ills—including exposure to different types of venom or other natural toxins.

To that end, two French scientists embarked upon a plan to study the effects of the Portuguese man-of-war's toxin on the body.[1] Paul Portier was a French physician, biologist, and physiologist with a

keen interest in marine biology.[2] Each summer, Portier joined Prince
Albert I of Monaco, who was also an avid ocean enthusiast, on the
Princesse Alice II, the royal's retrofitted yacht. The prince had trans-
formed the luxury yacht into a modern scientific research vessel,
equipped with the latest laboratory equipment and a full research
staff. Albert I and his scientific director had noticed that despite the
fragility of the Portuguese man-of-war's tentacles, fish that merely
brushed against them became instantly trapped. Sailors that came
into contact with them experienced debilitating pain and sometimes
fainted. The prince suspected that the man-of-war was capable of
producing a highly potent poison and asked Portier to investigate. In
the summer of 1901, Portier invited his colleague at the Paris Faculty
of Medicine, Dr. Charles Richet, to join him aboard the *Princesse
Alice II* to study the effects of the class of contact toxins produced by
Portuguese man-of-wars, other jellyfishes, corals, and sea anemones.

Charles Richet was a physiologist, like Portier, and the eccentric
son of a famed surgeon. In his youth, Richet had wanted to become a
writer and even had two plays produced in Paris, but his father ultimately forced him to go into the family business—medicine. Yet even
after becoming a physician, Richet maintained an interest in not only
the literary arts but also subjects as varied as paranormal phenomena,
socialism, and pacifism.[3] In 1890, Richet even built an airplane to
further his personal curiosity about aviation. Charles Richet's interests within the field of physiology were just as diverse, and he pursued
them with as much enthusiasm. In July 1901, it was ultimately Richet's
interest in toxins that brought him aboard the *Princesse Alice II.* But
it would be his stubborn pursuit of whatever sparked his interest that
would be an invaluable asset to the study of immunology.

Portier and Richet's original plan to study the man-of-war was
simple. First, they would methodically extract samples of tissue from
the marine animal's various body parts. (In actuality, the Portuguese
man-of-war, or *Physalia physalis,* is a symbiotic organism made up of
four distinct polyps that function together as one.) Then, they would
grind up those tissue samples and add them to a basic solution of sand
and seawater, injecting the solution directly into animals—in this

case, pigeons and guinea pigs brought aboard in ample supply for this very purpose. Eventually, they hoped they would discover exactly which parts of the man-of-war produced the paralyzing toxin and better understand the basic biological responses behind the paralysis it induced, leading to a better comprehension of both the method of delivery and the toxin's deadlier physical effects.

In the middle of conducting this research, however, Portier and Richet became suspicious that lab animals repeatedly subjected to diluted amounts of the toxin could develop a tolerance to it. Given enough time between injections, and the right amount of toxin in each solution, they theorized that the animals might become completely immune to the effects of the man-of-war's toxin.

That fall, having returned to Paris, Portier and Richet set up a series of experiments to test their hypothesis. The Portuguese man-of-war is only found in tropical waters, however, and was prohibitively expensive to import into their laboratory in the city. They decided to use the toxin of a common genus of sea anemones, *Actinia,* instead.

First, they injected differing amounts of sea anemone toxin into a few dogs and noted each dosage's effects. On the *Princesse Alice II,* Richet had become fascinated by each individual animal's varied reactions to the same toxins, which were carefully noted and tracked. As a physiologist, Richet had surmised that something about each animal's unique physiology, or its individual characteristics, had an effect on its biological reactions. In Paris, Portier and Richet became well acquainted with the personalities and quirks of their canine subjects, which made it far easier to track their idiosyncrasies. Some dogs in the lab (those receiving smaller doses) sickened and developed a rash at the site of injection; others (those receiving larger doses) died a few days after the injection. If any of the dogs remained relatively healthy after a diluted dose of toxin, Richet and Portier waited a period of time and then injected them again, hoping to trigger the dogs' natural immunity.

During these initial experiments, their favorite dog, Neptune, was given a low dose of sea anemone toxin and remained healthy. Three days later, Portier injected another small dose without discernible reaction. To maximize Neptune's chance at developing immunity, Portier and Richet decided to wait three full weeks to give the dog an additional injection, giving his body what they believed would be enough time to build up a stronger tolerance to the toxin. What happened next would change the course of immunological science and the way we think about our immune system's basic functions.

Seconds after Portier injected Neptune with the third and final dose, the dog began to wheeze. Quickly unable even to stand, Neptune lay on his side, vomiting blood, his body convulsing. He died in just twenty-five minutes. When Portier informed Richet of Neptune's death, Richet realized that instead of becoming immune to the toxin, Neptune had instead become *more* sensitive. The reaction both saddened and puzzled him. It went against the reigning paradigm of germ theory (that the immune system was only about defending against outside invaders and that priming it would induce immunity), but reinvigorated Richet's ideas that biological individuality was a subject worthy of more scientific study. Why, Richet pondered, did some dogs tolerate the toxin better than others? Why had repeated, spaced-out, minuscule doses killed Neptune? Was Neptune's reaction an individual quirk of his unique biology, or just a generic—and hence repeatable—bodily reaction? And, more important, could they learn to predict or even induce these terrible reactions in dogs or other animals in the laboratory?

Over the next few years, Richet continued to experiment with different poisons in his lab in Paris, only this time he was trying to induce negative reactions similar to Neptune's. Eventually, Richet learned to produce at will what he called "hypersensitivity"—or an elevated reaction—in dogs, rabbits, and guinea pigs. Through repeated inoculations, Richet's lab animals became more, not less, sensitized to the toxins (largely because some of our immune cells are capable of "remembering" things we've been exposed to in the past in order to help mount a stronger response to them during subsequent encounters). If immunity meant defense against foreign substances, then what happened to Neptune and Richet's lab animals was its opposite. For Richet, hypersensitivity was the antithesis of immunity rather than a part of the same basic immune defense function gone haywire. If immunity was about developing a natural protection or defense against microscopic bodily invaders, with the immune system helping the body, then the hypersensitivity that Richet was seeing in his animals was about developing an overreaction against foreign substances—one that might have been trying to aid the body but ended up harming it instead. Thus, he named the reaction anaphylaxis, or "backward defense."

After several more years of research, Richet mused that perhaps anaphylaxis was a beneficial reaction against some short-acting toxins and was ultimately part of an intentional response system that could easily backfire, causing severe illness or death. In 1913, Richet won the Nobel Prize in Physiology or Medicine for his work on anaphylaxis. In his acceptance speech, he posited that immunity and anaphylaxis were both examples of what he called "humoral personality." For Richet, any creature's individual characteristics determined how its body would respond to the introduction of something like sea anemone toxin. Richet argued that while every animal had a similar immune system, made up of similar parts, no two would ever respond exactly alike. It was important, he stressed, to study why some individuals responded so poorly.

A FAMILY HISTORY OF INDIVIDUALITY

The idea that someone's physical or mental quirks might play a role in disease causation was not new, even in 1901. Physicians had been fascinated by variations in patient responses to illnesses, as well as the techniques and tinctures used to treat them, for centuries. A patient's bodily constitution and individual temperament were seen as highly idiosyncratic and thus were always considered during the diagnosis and treatment process.

Doctors from the nineteenth to the early twentieth century took detailed notes not just about their patients' physical conditions and self-reported symptoms but also about their observable mental and emotional states. Most physicians referred to these variations as part of the natural "idiosyncrasies" of biology. An idiosyncrasy included any abnormal reaction of an otherwise normal person—anything not caused by the typical observed progression of an illness. Idiosyncrasies were thought of as "functional aberrations"[4] and were the bane of the medical profession. Idiosyncrasies meant that symptoms were often difficult to categorize, making diagnosis challenging and treatments nearly impossible to standardize. Humans, doctors at the time lamented, were unlike Richet's lab animals; they weren't as easy to experiment on or to manipulate, making scientific discovery on human immune reactions slower and rife with obstacles.

Richet's discovery of anaphylaxis, however, did seem to dovetail with what was then known about the medical condition called "hay fever" or "summer catarrh." During a lecture in London in 1881, Dr. Jonathan Hutchinson described cases of hay fever as "individuality run mad."[5] Prior to the discovery of allergic response by Clemens von Pirquet in 1906, physicians believed that their patients' respiratory troubles were largely caused by having a nervous temperament, and hay fever was considered a disorder of the nervous system, not the immune system. Physical and mental "sensitivities" were thought to run in families. Despite research conducted in the late 1880s showing that exposure to pollen directly induced respiratory attacks (we'll revisit this discovery in chapter 5), most doctors continued to think that their patients' negative physiological reactions to pollen could not be the sum cause of hay fever. There had to be *something* more to it than biological responses, since it was clear that while some individuals exposed to pollen sneezed and had bronchospasms, others did not. In addition, many hay fever patients continued to have attacks throughout the year—not just during the pollen seasons. The missing element, physicians surmised, must be related to the enervated nervous systems of severely allergic patients because pollen was otherwise innocuous. Hay fever sufferers had likely inherited nervous dispositions that especially primed them to their bouts of asthma.

By the twentieth century, it was common knowledge that allergies ran in families, as had been demonstrated time and time again through the taking of detailed family medical histories,[6] a process that was (and still is) an effective way to uncover inherited conditions. Since hay fever and asthma were considered hereditary, a patient's entire family history of hypersensitivity became a crucial part of their diagnosis. Dr. William S. Thomas, an allergist working in New York City in the 1920s and '30s, commonly asked his patients if their immediate blood relatives had ever suffered from asthma, hay fever, hives, food idiosyncrasy, migraine, eczema, arthritis, rheumatism, and "coryza"—a term for year-round hay fever or "sniffles."[7] Allergists would then use patients' responses to construct detailed family charts, going backward and forward at least one generation, two whenever possible. At the center of the subsequent chart was the patient, connected by solid lines to his parents and his children.

In the case of patient Y, detailed in an early medical text on allergy, Y's father, X, born in 1778, had been so allergic to cream and eggs that legend had it he had been poisoned by a meringue (a light and fluffy dessert made from whipped egg whites and sugar). Like his father, Y, born in 1807, was also highly sensitive to cream and eggs. Y's second son was intolerant to eggs, his eldest daughter to both eggs and cream, and his youngest daughter to eggs. Of Y's four children, only one was entirely unaffected by allergy. Of Y's grandchildren, only one—his granddaughter by his eldest daughter—had inherited his intolerance to eggs. According to the prevailing thinking of the time, Y's granddaughter could "blame" her great-grandfather X for her affliction. But puzzles remained: Why did one sibling show signs of their grandparents' or parents' allergy while the other exhibited none? Why did the child of a parent with asthma develop eczema instead? The propensity to allergic conditions was thought to be completely biological, or inherited, but the form it took was idiosyncratic, or malleable and unpredictable.[8] It seemed obvious at the time that genetics—or inheritance—played an outsize role in the causation of all allergies, even if the specific biological mechanisms remained shrouded in the body's many mysteries.

Writing on idiosyncrasies in 1927, Sir Humphry Rolleston, general physician to King George V, argued that the "inborn" nature of hypersensitivity was obvious: "In the same family different manifestations may be seen in brothers and sisters, and more than one form may occur in the same individual. When there is a bilateral heredity the incidence is higher than when it is unilateral only."[9] In other words, the more people in your immediate family who had allergies, the more likely you were to express some form of allergy yourself.

In the 1930s, the renowned allergists Arthur Coca and Robert Cooke labeled this inherited sensitivity "atopy" to try to distinguish it from Richet's anaphylaxis, which they argued was an acquired condition and not inherited like asthma or hay fever. A 1932 book on allergy suggested that anaphylaxis was evidence of the body's "inborn resentment" against components of its environment.[10] It quoted Richet's own description to argue that anaphylaxis could be thought of as "the last stand of the race against adulteration." Early allergy textbooks described anaphylaxis as "usually acquired" and allergy as "frequently inherited." When it was inherited, anaphylaxis was thought to be passed down only from the mother and caused only by the same substance as the mother's allergy.[11] And while children seemed to outgrow anaphylaxis (something that we now know is more likely to happen with some allergies such as egg allergy, but less likely to happen with others like peanut or tree nut allergy), allergy was a lifelong problem. Allergic reactions were also thought to be much more idiosyncratic than anaphylaxis, which seemed to be more predictable in its manifestations.[12] Writing in 1931, Coca felt that asthma and allergy were overwhelmingly hereditary in nature, but the environment clearly contained inciting factors like pollen.[13] He thought allergies must be caused by "reagins" in the blood, or sensitizing agents particular to each individual, predetermined by their genes. Ultimately, Coca argued that pollen, weeds, and automobiles might trigger an allergic attack, but only in patients who were genetically susceptible in the first place.

Eugenic theories—which were rife in American medicine until after the conclusion of World War II—contributed to the study of genetic differences in allergy rates by race. Doctors in the United States reported that Native Americans did not have allergies, despite living in the same environments as whites who did (including some of the doctors working on reservations in Arizona, Wisconsin, and South Dakota). Only Europeans and Americans of European descent—read "whites"—were thought to be capable of experiencing allergic responses. Allergy patients were regularly categorized as hailing from white, urbanized, and wealthy families. One allergy pamphlet put it like this: "As one would expect, one finds allergy most commonly in highly sensitive, well-educated persons, and in their children. One expects this because a person usually has to be generally and nervously sensitive in order to be sensitive to dusts and pollens."[14] The long history of associating allergies with particular races and personality types lingered, even as research proved the ubiquity of allergic responses across all races, genders, and social classes.

But as the sciences of immunology and genetics both advanced, the thinking about genetic inheritance began to shift. By the 1950s, a pamphlet published by the Allergy Foundation of America assured the public that allergies were *not* inherited.[15] What *was* inherited was a genetic tendency to develop allergies, but not inevitably, and not necessarily the same allergies suffered by grandparents, parents, or siblings. Today we know that every allergy is unique—or idiosyncratic—to each person. And while we know with absolute certainty that family history matters, what remains unclear is just how big a role DNA really plays in the development of allergies in both children and adults.

THE GENETICS OF ALLERGY

Let's be perfectly clear about one thing from the outset: There is no single gene, gene segment, or region of our DNA that causes allergy.

All too often, when we search for the underlying biological cause of a disease, what we really want is a proverbial smoking gun. We want something specific and definite—and preferably something we can alter, manipulate, or fix. But allergy causation is not that simple, biologically speaking. And although allergies often run in families, the genetics behind their development are anything but straightforward. Even the basic cell biology behind allergic reactions—driven in part by our genetics—is not well understood. As Dr. Dean Metcalfe, a researcher working on mast cells at the National Institutes of Health, told me, "The mechanisms behind allergy are very complex and we're really far behind."

Researchers doing basic science in the field of allergy often turn to genetic mining for clues. Allergy patients' DNA is collected, stored, and sequenced. The resulting data sets can then be compared and checked against nonallergic people's DNA to spot any significant similarities or differences. Gene segments that are shared by many allergic people might help us home in on the biological mechanisms driving our overreactive immune responses. Scientific researchers trying to understand the basic functions of our immune cells, like Metcalfe, are interested in genetic studies because clues found in our DNA may eventually lead to the development of better diagnostics or treatments. The hope is that by finding the gene segments associated with a higher prevalence of allergies, we might be able to either prevent them from ever developing in the first place or interrupt the biological pathways that drive harmful immune responses.

But when it comes to the genetics of allergy, it's important to stress that correlation is not necessarily causation. Although the researchers I spoke to were all uniformly agreed that our genes likely play a pivotal role in the development of our allergies, they were also quick to point out that our genes are not entirely to blame for them either. The human genome comprises approximately thirty thousand genes. Each of those genes interacts with both other coding (genes) and noncoding segments of our genome, as well as our larger environment, to regulate all our biological functions, including our immune system responses. So of course genes are involved in allergic reactions; that much is a given. The larger question at stake here has to do with just how much direct influence our genetics may have over the expression of allergies throughout our lives.

Genes can be affected by a variety of factors: hormone levels,[16] our age,[17] or things within our immediate environment[18] (like plastics, as we'll see in chapter 5). Genes also interact with one another, affecting their expression in complicated ways. Part of the difficulty in pinning down the genetic causes of allergy resides in how many different genes might be involved in producing allergic-type immune responses. In a recent study using genetic data from more than 350,000 participants, researchers found 141 different regions on the human genome that correlate to an elevated risk for developing hay fever, asthma, and eczema.[19] The trouble is zeroing in on which specific genes control for which parts of our immune systems and how.

The Case for Genes: The Barrier Hypothesis

I first heard about "the barrier hypothesis" when I visited Chicago. It was early fall and the city's flower beds around my hotel downtown were stuffed with colorful mums and pumpkins. As I made my way to a café nestled near the campus of the University of Chicago to meet one of the country's top eczema experts, I wondered how much the city's penchant for beautifying its streets contributed to its citizens' hay fever woes.

Dr. Peter Lio is a clinical assistant professor in the Department of Dermatology and the Department of Pediatrics at Northwestern University and a practicing dermatologist who specializes in eczema care. He's an affable, gregarious man in his early forties who understands and empathizes with those who suffer from severe atopic dermatitis, or eczema. He co-founded the Chicago Integrative Eczema Center, renowned for its holistic approach to treatment, and is currently its director. Lio is a little late to our interview—one that he generously agreed to squeeze in after a long day of seeing patients and before picking his daughter up from school. We're sitting outside at a wooden table, and bees are constantly buzzing around us, so much so that we sometimes need to pause our conversation and duck our heads to avoid their flight patterns. Lio admits he's not a fan of bees either, but the café is jammed, so outside it is.

Our conversation begins with Lio explaining the dramatic impact of skin allergy and eczema on patients. Traditionally speaking, ec-

zema has not been categorized as an allergic disorder, but that viewpoint is slowly changing. Lio explains that "eczema" is a bad term for
a complicated set of symptoms and triggers. Not everyone who has
eczema will be triggered by allergens (some can be set off by changes
in temperature or activities like exercise), but the skin's reaction during an outbreak of eczema (no matter what its trigger was) is similar
to other allergies in that it involves the immune system. Eczema can
be fairly debilitating in moderate to severe cases, and because of Lio's
reputation in his field, he often sees patients who are, as he tells me,
at the end of their ropes. Often, they've been trying to figure out their
condition and its trigger for years. By the time they visit the center,
they're exhausted and frustrated. Diagnosis of the condition is difficult and treatment—relying mostly on dangerous topical steroid
creams—is often ineffective. Still, Lio is upbeat about what the future
holds, thanks in part to recent scientific discoveries.

"The big break came about ten years ago," Lio explains, "when we
discovered that atopic dermatitis was connected to a mutation in a
gene called *FLG,* encoding a protein called filaggrin."

A longitudinal study of a cohort of more than two thousand pregnant women[20] in England and Scotland was conducted by Dr. Mukhopadhyay at Brighton and Sussex Medical School in the United
Kingdom. The research team collected cord blood for genetic sequencing and followed up with mothers to ask about allergic conditions in their children at the ages of six months, one year, and two
years. Mukhopadhyay and his team found that a common gene defect
that affects the production of the skin protein filaggrin is linked to the
development of eczema, wheeze, and nasal blockages in babies as
young as six months old. This suggests that babies born with this genetic variation might be more susceptible to developing allergic conditions from birth. The skin barrier theory of allergy causation posits
that skin defects that cause the skin to be more porous in early age
allow allergens (and possibly other foreign materials) to pass through
the skin barrier and into the bloodstream, triggering immune cell responses. Lio is excited about the *FLG* gene mutation's link to eczema

because he finally has something he can tell his patients to explain what is happening to them. Between 15 and 20 percent of patients with atopic dermatitis carry this gene mutation.

"For the first time, you can look at a patient and say, 'I know why you have this disease, because you're missing this gene, so you have leaky skin,'" Lio says. "And that's pretty deep, right? We crossed into new territory, because for the first time we could actually give an answer. This concept of 'leaky skin' has been really fruitful because now we know that's how allergens, irritants, and pathogens get in and probably explains the skin microbiome abnormalities in these patients."

Here's how Mukhopadhyay describes the importance of the filaggrin discovery: Picture your skin as layers of paper tightly stapled together. To demonstrate, Mukhopadhyay holds up his hands, which are placed flat and horizontal to each other. The fingers of one hand slightly overlap the fingers on top of the other hand.

"You've got layers of keratin sheets, which are lying like this," he says. "One after another, in sheets."

The filaggrin proteins act like staples, keeping the layers of your skin tightly coupled together. Filaggrin, in essence, produces firm skin with a healthy, interwoven barrier. A healthy skin barrier keeps things outside your body from seeping inside it. The problem is that the staple doesn't work well in 10–15 percent of infants.

"In these babies, the allergens will be able to get into the body much more easily than they would if the staples were working nicely and the filaments held tightly," Mukhopadhyay explains.

In other words, the lack of filaggrin, due to a genetic mutation, causes the "leaky" skin that Lio described. According to Mukhopadhyay, this mutation is widely present in the population at large and has been for an exceptionally long time.

"Filaggrin defects were definitely there in our genome, five thousand years ago, three thousand years ago," Mukhopadhyay tells me. "But where was the house dust mite? Now we huddle in soft, padding-clad sofas to watch television in damp, warm environments and breathe in house dust mite feces. We allow our soft mattresses to smear dust mite feces on our bodies at night when we sleep. And the immune system is perhaps coping okay with exposure to dust mites in three out of four people. They've got one or two or three or five or ten little tweaks in their genetic makeup that's making all the difference—and we are not understanding those tweaks. We are just prescribing steroids to the one person whose immune system isn't coping instead of trying to understand the three people whose immune systems are."

In the case of *FLG* mutations, a specific gene sequence might be driving an allergic response. This means that researchers might be able to figure out a way to fix the skin barrier, stop the skin from "leaking" allergens into the body, and prevent eczema from developing in the first place. From Lio's perspective, this is evidence for why we should start thinking about allergy patients according to their genetic subtypes. Treatments that work on the 20 percent of patients with leaky skin caused by *FLG* mutations might not work as well on the 80 percent of patients who have an intact skin barrier but a different underlying biological mechanism driving their skin irritation. Genetically speaking, then, eczema isn't a single skin disorder; it's multiple skin disorders with similar symptoms.

Collecting more patient DNA samples and doing more genetic mining might help us find more similarities in the genetic makeup of allergy sufferers that in turn lead to better treatments. Mukhopadhyay has a theory: If we could use precision medicine to test which babies have the *FLG* mutations, then we might be able to block the entry of allergens and prevent allergies from developing, perhaps by somehow making leaky skin less permeable. A randomized study that tested the use of skin emollients, or moisturizers, on all babies would show a success rate of only 15 percent because only 15 percent of them would have the filaggrin gene mutations. The study results would suggest that skin emollients aren't effective in preventing the development of eczema in infants. But Mukhopadhyay argues that if we ran a similar study treating only the 15 percent of babies with skin barrier defects such as those resulting from filaggrin gene mutations, we might see that emollients—or similar creams that strengthen and enhance the skin barrier—work very well to prevent eczema. For Mukhopadhyay, this is the true promise of genetics and studying gene-environment interactions: Knowing which underlying genes correlate to the development of allergy might help us lower rates of allergy in the future. What's more, he thinks he has definitive proof that this approach can work.

Enter the adorable house cat.

Once Mukhopadhyay's research team found the correlation between the *FLG* gene mutations and higher rates of asthma, respiratory allergy, and eczema, they began to wonder what would happen if a baby with *FLG* mutations grew up with a cat—and thus copious amounts of cat dander that might more easily permeate their skin barrier—in their household. What would be the risk of that baby developing atopic dermatitis by the age of two?

Mukhopadhyay's research team designed a study to find out.[21] They enrolled babies both with and without *FLG* mutations, and with and without pet cats, in their study. They discovered that babies born without any of the common *FLG* mutations and who lived without a cat still sometimes developed atopic dermatitis—but at the low rate of 10–15 percent. If the baby didn't have the filaggrin gene mutations and the family lived with a cat, the rate of eczema went up slightly. If the baby carried the *FLG* mutations and the household didn't have a cat, the baseline rate went up substantially, to 20–40 percent. And the rate soared if the baby had the *FLG* mutations *and* lived in a household with a pet cat. More than 95 percent went on to develop eczema.

The take-home here, Mukhopadhyay argues, is that by genotyping children at birth, we might prevent some childhood cases of eczema[22] simply by alerting parents to the potential dangers of owning a cat.[23] The parents could then alter the home environment to avoid a poor interaction with their children's genes. In that way, precision medicine is a return to a time when physicians would treat individual cases of allergy by taking patients' home environments into consideration.

"Genetics is nothing new," Mukhopadhyay suggests. "Personalized genetic phenotyping and genotyping are making us come back full circle to our ancient practices, but it's empowering us to look at idiosyncrasies in a much more scientific manner. And this way of thinking is only in its infancy. In fifty years, people will be able to make choices about their environments and lifestyles in a much more deliberate way, based on their genetics. This is the future of allergy medicine—of all medicine."

The Case Against Genes

Immunology researchers at the National Institutes of Health have been working on the problem of allergic immune responses for years. What they have found is, that while genes definitely play a role in the development of allergies, they don't tell us the entire story. I first met Dr. Joshua Milner while he was still a physician-scientist at the NIH campus in Bethesda, Maryland. Today he is chief of the Division of Allergy, Immunology, and Rheumatology in the Department of Pediatrics at Columbia University and professor of pediatrics in the Institute for Genomic Medicine at Columbia University Irving Medical Center. Milner is famous not only for his work on the genetic pathways of allergic immune responses but also for his research linking immunodeficiency diseases with allergies.[24] When it comes to understanding how genes correlate with allergic disease, Milner is one of the best people to go to.

On a crisp winter day, I sat down with Milner in his NIH office to discuss his groundbreaking genome-wide association studies looking for new biological pathways that lead to allergic responses in humans. Milner is a fast talker. Scribbling notes as quickly as I could, I was still barely able to keep up. We covered a lot of ground that afternoon, but the thing that struck me the most about our conversation is this: The genetic component of allergy can show you who *might* be at greater risk but not who will *definitely* develop allergy.

To illustrate this, Milner told me about a gene called *MALT1* that has been associated with a greater risk of peanut allergy if the children who carry the gene are exposed to peanut proteins late in their development (generally speaking, after two to three years of age). Yet infants with the exact same *MALT1* mutation who are given peanuts much earlier in their childhood development had a tenfold protection against peanut allergy. In other words, the same gene is both protective and not protective; it all depends on the timing of the child's exposure to peanuts. The gene-environment interaction, Milner explained, is key here, not the gene itself.

After visiting the NIH campus, I traveled to the world-famous Cincinnati Children's Hospital, to talk with one of Milner's close friends and colleagues, Dr. Marc Rothenberg, who is the leading expert on a condition called eosinophilic esophagitis (EoE). EoE is a rare allergic disorder of the esophagus. Patients with EoE have an accumulation of eosinophils, a type of white blood cell involved in immune function, in their esophageal tract. Symptoms are dreadful and treatment is difficult. Some patients are allergic to so many different food groups that they suffer nutritional deficits as a result of dietary restrictions. When I asked Rothenberg whether or not genetics is a cause of EoE, he demurred. His lab sequenced and compared DNA from many families and found few genetic similarities between the families even though genetics was having an effect.

"There was very low sharing of genetics between one family and another," he explained. "What that indicates is that there's a lot of heterogeneity in the genetic basis of the disease. Most of the patients that we see with allergies have substantial involvement of the environment interacting with the genome, influencing disease susceptibility and phenotype. The mechanism involves epigenetic changes in gene expression in a variety of cells including immunocytes."

Rothenberg pointed to the results of studies looking at the rate of allergies in twins as further evidence that genes aren't the main source of allergic disease. In a study of both fraternal and identical twins conducted by the Elliot and Roslyn Jaffe Food Allergy Institute at the Icahn School of Medicine at Mount Sinai, only 66 percent of identical twins, who share the exact same DNA, shared a peanut allergy.[25] Fraternal twins, who don't share the exact same genetic code, had food allergy in common 70 percent of the time. It's obvious to Rothenberg that since we don't find 100 percent consistency in who develops an allergy in sets of identical twins that *it is not the DNA but the shared environment of the siblings* that is driving the allergic response.

"The DNA is clearly contributing," Rothenberg said, "but it's actually not the main factor."

Ultimately, that's good news, Rothenberg reminded me. For the most part, our DNA is not changeable. We cannot alter our DNA to control allergy, but we might be able to alter our environments.

Rothenberg regularly collects biopsies from inflamed tissue in the esophagus and stores them for further study. "We have over thirty thousand samples in my lab that are derived from allergy patients, including the inflamed tissue derived from gastrointestinal biopsies," he said. "And that allows us to probe that human allergic information for the first time in a very high level of inquiry." He explained that studying an extreme allergy phenotype like EoE genetically might allow scientists to pick up a higher signal in the data (as opposed to studies that look at more common allergic diseases that have a lower signal-to-noise ratio). The fewer people who have a severe disease, the more clearly researchers can spot any similarities in their DNA that people without the condition don't have. That data can then be used to help uncover the more common allergic pathways involved in less extreme phenotypes of allergy—asthma, eczema, hay fever—since similar biological mechanisms are likely at play in those allergic conditions as well. That's exactly what Rothenberg's lab at Cincinnati Children's is busy trying to do.

Back at the NIH, Milner shows me his massive cryogenic tank— a large cylindrical steel machine that houses thousands of patient blood samples. It's a treasure trove of future knowledge, but it will take time to explore. As I am getting ready to leave, Milner stresses to me that it's important to understand that the same genes perform multiple actions—some of which have nothing to do with immune function. He tells me that when mice are missing the gene for IL-4, a protein inhibitor and one of the biological components involved in allergic response, the mice become more forgetful.[26] So it's possible, Milner suggests, that one of the same genes that drives allergic response in humans may also be involved in brain processes that are foundational to memory.

"How many nerds at MIT did I go to school with who had terrible allergies?" Milner jokes. His answer? Most of them.

A Genetics Curveball

On a spring day in 2019, I rented a car and drove from New York City to Ithaca, New York, to meet with Dr. Avery August, a professor of immunology at Cornell University. August conducts research on immune cell function, specifically on one of the powerhouses of immune response—the class of white blood cells known as T cells. T cells roam our bodies looking for foreign particles; their "job" is to make decisions about any antigens that they encounter. In other words, T cells are those "curators" of the human body that August talked about earlier in the book. They help to decide which things can and cannot be a part of us.

August's office, tucked away in one of Cornell's newest science buildings, is neat and organized. During our meeting, he's somehow both relaxed and alert. He loves immunology—it's clearly not just his job, it's his calling. When I ask him what is driving the recent increase in the rate of allergies, August explains that it can't be genetics altering immune system function.

"Genetic change is much slower than environmental change," he says. "There's always been genetic change in our immune systems as we find ourselves in different environments. But it takes a very long time to change."

He tells me about the laboratory mice extensively used in immunological studies. They are not "normal," run-of-the-mill mice like the ones you might spot in your kitchen; mice used in immunological research studies have genetic diversity that is highly controlled.

"These are *exactly* the same mice, genetically speaking," August explains to me. "They're genetically inbred and so there's no change in their DNA. Their interaction with their environment is the only change."

When scientists produce an allergic reaction in laboratory mice, they generally do so by altering components of the mice's diet or environment. Sometimes laboratory researchers trying to understand things like mast cell function or histamine response might special-order mice with specific gene segments knocked out. August underlines that genetics is obviously important—we all react to the same stimulus slightly differently on the basis of our genetic blueprints—but when environmental changes are added on top of those genetic differences, we observe significantly different responses to the same allergic triggers. This means that there is nothing necessarily "wrong" with the genetics of an allergic individual. DNA is not the fundamental problem when it comes to immune system function—environmental triggers are. In fact, the immune system of an allergic person is functioning exactly the way in which it was designed.

August has another argument against seeing genetics as the smoking gun behind the recent rise in allergy rates: our own cells. He has spent most of his career trying to understand why T cells react in the ways they do. All of the T cells contained within our bodies are, of course, genetically identical; nearly every cell in the body contains the same DNA. What's more, because they all exist inside us, they have exactly the same environmental exposures, too. Everything that happens to us happens to them. If genetics could truly predict allergic response, then all of the cells in a single person should respond exactly alike. The problem is . . . they do not.

"I've spent a lot of my time trying to understand how the cell, when it first comes in contact with that sensitizing antigen, decides to do one thing or the other," August says.

He uses a cup of water on the table and his hands to illustrate. One cell, his left hand, runs into the cup—an antigen—and notices that it is not supposed to be there. It has to make a decision about the cup. Is it good or is it harmful in some way? Can it stay or does the cell need to alert its neighboring cells that there is a problem? Another cell, his right hand, encounters the same cup and has to make the same decision. What August's research group has discovered is that individual cells from the same body will make different choices in this moment. Some will move on, effectively allowing the cup to keep rest-ing on the table. Some will decide that the cup needs to be removed immediately.

"Using genetic tools, we can mark the cells that are reacting," Au-gust says. "We can also tell which ones aren't, and so now we're com-paring these two populations. Why did this cell react, and these other cells not react? Is there something about these two different states of the cell that tells us about how to prevent one from happening?"

Using high-throughput single-cell RNA sequencing of large popula-tions of cells, researchers at MIT were able to identify T cells that pro-duced inflammation in food allergy patients who react to peanuts.[27] They are also studying the same T cells to see if they respond differ-ently after the patient has completed immunotherapy. The sequencing technique, which can capture messenger RNA, allows researchers to see which genes are being expressed at any one time—giving them a better understanding of cell functions. An individual T cell's RNA is given a barcode so researchers can track which T cells target the pea-nut antigens, hopefully leading us to a better understanding of how T cells make decisions about how to react.

Our immune systems are among the fastest evolving of our bio-logical systems. August suggests that they have to be because our im-mune function is literally a matter of life and death. But even so, our immune cells can't keep up with the pace of human-altered environ-mental change. Understanding how our DNA contributes to allergies is perhaps less important than understanding how our basic immune cells make decisions about what they come into contact with in the first place.

But perhaps the most interesting question about the relationship between our genes and the development of allergies—whether or not that connection is critical or incidental—is this: Why would our own cells have the ability to harm, or even kill, us in the first place? If my half of my father's DNA contains the blueprints for a similar reaction to bee stings, then why would evolution select for that response and pass it on to me? Or, as Dr. Alkis Togias, the branch chief of Allergy,

Asthma, and Airway Biology at the NIH, put the question to me, "Why would the development of the immune system lead to such a problem, which seems to go against nature?"

THE TOXIN HYPOTHESIS

Dr. Steve Galli had a hunch about the evolutionary underpinnings of allergy. As a pathologist and an immunologist who researches mast cells and basophils at Stanford University, Galli, like many of his colleagues working on the basic cell science behind allergic responses, wondered if our ability to react to something like sea anemone toxin might have been useful at some point in our evolutionary past. An overreaction to something harmless like peanut protein or dust mites today could be a relic of a much older part of our immune system—one that had evolved to deal with something quite different, and likely more dangerous. Perhaps allergic immune response, which is largely dysfunctional in the twenty-first century, had given our ancient ancestors a survival advantage.

"I was interested in this apparent paradox that you've got an extremely active form of immunity that can be triggered, essentially immediately, and can have a catastrophic ending, like your father," Galli explained to me. "Why would evolution come up with that? It seems so maladaptive. So why do we have these types of activity?"

Galli's answer relates back to the differences between our innate versus our adaptive immune system. To recap, our innate immune system is online from the moment of our birth. It is, if you will, our body's first line of defense. Mast cells, basophils, and eosinophils—some of the immune cells that contribute to allergies—are part of our innate immune response. Innate immune responses are generic, meaning that they can respond to anything foreign that enters the body. Adaptive immune responses, on the other hand, are more specific. Immune cells like B cells or T cells learn what antigens, or foreign substances, to respond to and then "remember" them, making future responses to those same things quicker and stronger. Our innate system can respond to any threat immediately; our adaptive system needs to learn what to react to and it can take repeated encounters to ramp up a response.

"What you need immediate reactivity for is something that must be avoided quickly," Galli said. "You have to know about it quickly in order to avoid it quickly. What sort of things are they? They're stings of venomous insects. They're things that if you ingested them would kill you. So you want to develop a very rapid reactivity to them so that you don't eat them, you spit them out."

Galli wondered what types of things that anaphylaxis, an immediate hyperactive immune response, might be beneficial for combating. What would a human living twenty thousand years ago come across that would necessitate such a drastic triggering of mast cells in the body? One possible answer: venomous snake bites. Another possible answer: venomous insect stings or bites.

Galli asked me to think back to the moment my father was in his car, struggling to breathe. He asks me to think about my dad's physical reaction in a different way.

"A lot of the fatal anaphylactic reactions not only occur due to bee stings or peanuts but are in people who can't become fully supine after getting the reaction," he explained. "The common story is they're in the cab of a truck and they're stuck, and they can't flatten out. Because if you flatten out you can deal with a much lower blood pressure."

It makes sense, from a biological perspective, to slow down the circulation of a toxin through the bloodstream and to trigger the body's defenses against it. Galli suspects this may have been one of the earliest tasks of mast cells; they might have been crucial to our survival in pre-modern times.

"Mast cells go back before the development of antibodies," Galli explained. "They are very ancient components of the immune system."

One study estimates that mast cells first emerged more than five hundred million years ago.[28] Mast cells are incredibly old then—evolutionarily speaking. IgE, the human antibody associated with most allergic reactions, on the other hand, is a relatively recent addition to our immune responses. IgE causes hyperacute reactions that are mast cell dependent. These can be protective if the threat is a serious one and needs to be avoided in order to escape death. Even a few hundred years ago, Galli hypothesized, an immediate and strong immune response might have been helpful.

"At some point in our natural history, this mechanism was beneficial," Galli said. "Over the last two hundred years or so, this benefit has become less important, but the immune system still reacts to potential threats in the same way. Only these potential threats, instead of being poisonous snakes, are just food that's been applied in the wrong place, and so forth. And that has caused confusion in the immune system."

Galli and his team were already aware of something called "the toxin hypothesis" of allergy causation. The toxin theory was Margie Profet's idea. Profet is a brilliant evolutionary biologist and MacArthur grant winner. Her original theory was that allergies might be the body's method for expelling toxins and carcinogens. This idea had also been tied to research that found lower rates of some cancers[29]—especially gliomas—in allergy sufferers. Galli credits Profet and another researcher, James Stebbings, as originators of the toxin hypothesis. According to Galli, Profet and Stebbings were the first researchers to suggest that mast cells don't just cause harmful reactions but can also be beneficial.

"Stebbings said that one hundred or two hundred years ago, about the time that the first cases of hay fever were reported, people and animals were heavily bitten by insects," Galli explained. "Stebbings thought that the rapid reaction of mast cells and IgE-dependent reactions to these bites would tell someone that they'd better get out of that area immediately. It's like an early warning system and it probably saved lives. But you can't do an experiment in humans to prove that."

Instead, Galli's lab set up a series of experiments to test the toxin hypothesis in mice. One of the triggers of an allergic or anaphylactic reaction is endothelin-1, a peptide secreted by endothelial cells that is chemically homologous to a toxin called sarafotoxin, found in the venom of the mole viper *Atractaspis engaddensis*. Galli and his lab team showed that mast cells could degrade endothelin-1 and render it less toxic to mice. He and his postdoctoral fellow, Martin Metz, then wondered if mast cells might also be protective against sarafotoxin. The first experiment the lab conducted, using a synthesized peptide identical to the one in the venom, showed promise. The mice injected with sarafotoxin reacted in the same way as those injected with endothelin-1.

"Their blood pressure dropped and if they got a high enough dose, they died," Galli explained. "As far as the mouse was concerned, it really didn't matter whether it was getting endogenous peptide or the snake equivalent of it."

But Galli wasn't content with testing just one of the components of the snake venom, since natural venom is an admixture of many different toxic substances. What Galli and his research team really needed was whole venom from a snake that would have been in the day-to-day environment of early humans. They needed real mole viper venom. The problem was where and how to get it.

"It's an Israeli snake and it's not widely distributed," Galli told me, "but I knew of an investigator in Israel who kept some of the snakes in his lab."

The Israeli scientist, Elazar Kochva, had retired, but he still had some venom on hand. He was willing to give it to Galli, but there was an obstacle: Galli had secured a U.S. government permit to bring the venom into the United States, but Professor Kochva told him it would be better not to apply to remove it from Israel. Kochva suggested that it would be better for Galli himself to try to bring the lyophilized (freeze-dried) venom back to the United States from Israel. So Galli decided to fly to Israel to pick up the mole viper venom and bring it back to his lab.

"Have you ever been to Israel?" Galli asked. "They do security in a very interesting way. They've got people who are trained in psychometrics, and they stand in front of you and look right into your eyes, and they ask you questions very rapidly."

Kochva gave Galli a vial of venom that had been freeze-dried. The vial and its contents could be kept at room temperature for a few days without losing any of the venom's chemical activity. Galli recalled that Kochva told him he should carry the vial through Israeli security in his pocket and let it make the airplane trip home in his carry-on luggage. At the security checkpoint at the airport, an Israeli officer asked Galli questions, completely unaware that Galli had deadly snake venom tucked away in his inside pants pocket.

"I was standing there, and he was asking me all sorts of questions," Galli said, laughing as he recalled how worried he was that the venom might be detected. "But he never asked me whether I had venom in my pocket."

Technically, Galli said, he hadn't lied. The plastic tube didn't trigger any of the detectors and Galli made it onto the plane and back to California, the vial intact and the venom still chemically active.

In the end, the mice reacted in the same way to the whole venom as they had to the isolated, synthesized toxin. Galli's lab eventually tested venoms from two other snakes, the western diamondback rattlesnake and the copperhead. All of them produced similar results: Mice that contained mast cells were substantially more resistant to the toxicity of the venoms than were mice that genetically lacked mast cells. Moreover, mast cells treated so that they lacked carboxypeptidase A (a substance that can partially degrade some of the venom's components), one of the cells' stored enzymes, failed to protect the mice from the venom. The research team wrote up their findings and sent their work off to the journal *Science,* where it was published.[30]

Galli's lab did another set of experiments with venom from a poisonous lizard, the Gila monster, and found that it, too, along with the venoms of two different scorpions, triggered an effective immune response, but this time involving a different mast cell protease. Here was additional evidence that our innate immune systems evolved to rapidly defend against a variety of toxins—from stings to bites. The responses that Galli's lab was studying, however, happened after initial exposures to the venoms. What would happen if the mice that survived an initial dosage were given a second or third dose? Galli and his team went looking for the classic signal of allergic responses—the activation of the IgE antibody.

"We found that if you are a mouse and you survived the first injection of either honeybee venom or a snake venom, you would develop an IgE response to the venom," Galli explained. "Then when you were injected three weeks later, the IgE response would produce a rapid reaction to the venom that actually had survival benefit. So, having an IgE response to the venom helped the mouse's survival; it didn't impair its survival."

Mice that had been exposed to a smaller amount of venom could survive a larger dose later; mice that hadn't been previously exposed were not so lucky. If mast cells and IgE are protective, they may have given us an important evolutionary advantage.[31] The only problem with the toxin theory and Galli's experiments is this: We are not mice.

"This is the dilemma," Galli said. "Certainly we're different than mice. You can't do in vivo studies in humans with venom. You can only test in vitro [outside a living organism]."

In the United States, there are only ten deaths per year from poisonous snakes.[32] Worldwide, however, that number spikes to around one hundred thousand, typically in developing nations. There are fewer deaths from venomous insects or other creatures like the Portuguese man-of-war. Overall, death from toxins or venoms is fairly rare, which suggests that changes to our environment have made those parts of our innate immune function designed to cope with them far less advantageous.

Now a researcher working in Berlin, Martin Metz, one of Galli's former postdoctoral fellows, has continued studying responses to venom. He has shown that human tryptase (a stored mast cell enzyme that breaks down proteins) can degrade snake venoms. That's further evidence in support of the toxin hypothesis.

"So, as best we can tell, it seems likely that humans are similar to mice in having an IgE- and mast-cell-dependent resistance to some venoms," Galli concluded.

I find myself convinced by the research of Galli and his students. It seems likely that we conserved this type of immune system response for a reason. And that reason is likely to be protection against something in our environments. The drawback, I suppose, is that because our environments have been rapidly changing, those of us with strong immune reactions are left with a host of new problems.

GENETIC INHERITANCE IN THE REAL WORLD:
THE TYPICAL ALLERGIC FAMILY

So how does all this information stack up when it comes to the real-world issue of understanding the impact of our DNA on our allergies? Are allergies inherited or not? And can we use our own or our relatives' allergy profiles to predict which allergies our children might develop?

The answer to these questions are: yes, and no. But let's take a quick look at my own family to parse this out a little more concretely.

As far as I know, most of my grandparents remained allergy-free for their entire lives. Only my grandmother on my mother's side had any allergies. In her late fifties, she developed an allergic reaction to penicillin. As I explained in the first chapter, drug allergy is not mediated by IgE antibodies. Instead, my grandmother's T cells likely would have remembered encountering penicillin and developed a sensitivity to it. It was the only allergic reaction she ever experienced, and it was easily avoidable. In other words, while my grandparents' genetic makeup likely contained segments that primed my parents to sensitivities, it didn't lead to any overwhelming allergies in their generation. As we'll see in chapter 5, the reason for this might be because their immune systems had the benefit of being "trained" in different environments while they were still children in the first few decades of the twentieth century (an era with a lot fewer added chemicals, pollutants, and plastics).

My mother had no allergies whatsoever. Her brothers were also allergy-free, but her older sister—my aunt Grace—developed the same penicillin allergy as my grandmother (at around the same age). My mother's two younger sisters had a different father, who suffered from hay fever and asthma. My aunt Patricia has hay fever and experiences outbreaks of hives and itchy skin. My aunt Gloria had a severe allergic reaction to bee venom that landed her in the emergency room. For the rest of her life, she tried to avoid bees and carried Benadryl in case she was stung again, eschewing paying for the more expensive EpiPen her doctor told her to carry. My half-brother (we had different fathers) had chronic obstructive pulmonary disorder (COPD) at a young age after multiple childhood lung infections and a few years of breathing in noxious exhaust in the U.S. Air Force.

And as for my dad: His harrowing experience with bee venom allergy began our journey.

My genetic inheritance is a mixed bag. There is clearly a family lineage, but it's not a direct one-to-one causation. But what about the bee venom allergy? If someone on both sides of my family was prone to it, does that mean I might be more likely to have it myself? Not necessarily, but maybe. Since my IgE levels are low and my skin and blood tests were all negative, there's no way to know in advance of getting stung by a bee if I have that sensitivity myself.

Joshua Milner explained to me that about 5 percent of the general population of Northern European descent has a genetic mutation—an extra copy of a gene—that gives them high tryptase levels, which can lead to many problems, among them anaphylaxis from a bee sting. Tryptase is a protein in mast cells and a marker used to track mast cell activation during allergic responses. According to Milner, families with high tryptase "have itching and flushing and belly pain" but no evidence of disease, allergic or otherwise. And that sounds a heck of a lot like some of the symptoms I have started to experience since my own allergy diagnosis: sensitive, itchy skin; reddening skin or flushing; mysterious abdominal pain that has no clear etiology, or cause.

Allergies are clearly part of my family's genetic inheritance, and my predisposition to hypersensitivity was likely passed on to me by my parents as part of my unique DNA. And while my family is exactly what a typical "allergic family" looks like, our DNA alone cannot entirely explain our allergies. All of my symptoms are caused by genetic, biological responses to specific environmental triggers, and yet the specific type of allergy (localized allergic rhinitis) and its severity (mild) is different from anyone else's in my family—and that is not at all unusual. The truth is that genetics can only tell us so much about our propensity to allergy, but it usually can't tell us what we really want to know: in my case, whether or not I've inherited a sensitivity to bee venom.

Genes + ? = Allergy

Since the early days of immunological research more than a century ago, genes have been posited as one of the main forces behind the rise of allergic conditions like hay fever, asthma, eczema, and food allergy. But as we've seen, genes cannot be the sole—or even the primary— cause of all our irritation and hypersensitivity. Our DNA obviously plays an important role in what is driving the increase in allergies, but it is not the proverbial smoking gun. In fact, asking whether or not allergies are inherited is no longer even the correct question.

"The question is, How did we get a generational change?" Dr. Cathryn Nagler, a top microbiome and food allergy researcher, asked. "Because that's really what it is. People will tell you, 'There's no history in my family of this. Nobody had this before.' From parents with no family history of allergy to kids that have life-threatening responses to a crumb. And it's real. . . . Your allergy can develop at any point in your life. It used to appear between the ages of two and five. Now we're getting a lot more adult-onset food allergy."

All of our immune systems are dealing with the same changing environment. Ultimately, as Avery August has argued, that means that the solution to allergies "isn't necessarily a biological solution, it's a collective solution about what we do about all these other things that are impacting the rise of allergies." Our genetics might predispose some of us to be more or less allergic, but our DNA is not the ultimate problem. "If you really look at subpopulations where allergy is increasing, they're telling us something about where we are now as a world," August said.

People with allergic conditions are the canaries in the environmental change coal mine.

CHAPTER 5

Nature Out of Whack

A TALE OF THREE CITIES

As I write these sentences, the air is crisp, and the blue sky is strafed by cirrus clouds. Birds are chirping in the branches of trees dotted with newly budding leaves. Daffodils and tulips have popped up in flower beds along the sidewalks. The grass is waking up from its winter sleep and turning a vibrant green. The parks are dotted with small groups of people enjoying the sun and one another's company. It's a perfect spring day.

Only, it's not. Not really. Not for people with respiratory allergies or asthma.

For them, invisible particles in the air are making it difficult to breathe today, causing attacks of sneezing and irritating their eyes, nose, and throat: not just the copious amounts of microscopic tree and grass pollen that are circulating in the air right now, floating around on the breeze, lightly coating outdoor tables or caked onto the exterior of cars and trucks like yellow dirt, but also the combination of dust, ozone, nitrogen dioxide, sulfur dioxide, and other particulate matter too small to see, even with a microscope. All the detritus of our modern civilization is constantly swirling around our bodies, being inhaled along with pollen deep into our lungs—and that pollution is especially concentrated in the urban air surrounding our cities. Even on beautiful winter days, when no pollen and mold spores are circulating, the air is still replete with things that are irritating to our immune systems.

Is all the pollution in the air we breathe making our allergies and asthma worse? Could environmental changes—to both our natural landscapes and the climate itself—over the past two hundred years be driving the dramatic recent increases in allergy rates around the globe? If, as we've already seen, our DNA isn't solely to blame, then could our environment be the key factor causing all our allergies?

In short, the answer is a resounding yes.

But frustratingly, I need to include "sort of" as a coda. Like with our genetics, changes to our natural environment—or the physical landscapes in which we reside—appear to be at least partially responsible not only for an increase in allergy rates but also for a worsening of our normal seasonal allergy symptoms. If you've felt as though your eyes were itchier, your nose was stuffier, or your sneezing fits have been getting worse over the past few years, you're probably correct. The reason likely has something to do with changes to the average pollen load (the amount of pollen in the air), the air quality itself (whether on average it is good, fair, or poor), and the indirect effects of climate change on everything from the number of mold spores to crop production to trapped heat to the circulation of air.

In this chapter, we'll examine some of the evidence scientific researchers have amassed showing that recent environmental changes are both overwhelming and confusing to our immune systems, helping to drive the increase in the global rate of all allergic conditions throughout the last century. We'll look at the past, present, and possible future of hay fever and asthma sufferers in three cities— Manchester, England; Cincinnati, Ohio; and Chandigarh, India— to examine the ways in which changes to the air we breathe correlate with an increased risk of developing allergic disease.

Physician researchers working on hay fever and asthma in the nineteenth century suspected that shifts in agricultural production and polluted urban environments were directly linked to the development of their patients' hypersensitivity, or allergy. These early scientific theories about environmental allergy causation eventually paved the way for what would become known, more than a century later, as "the hygiene hypothesis." The hygiene hypothesis posits that changes to our environments—specifically a lack of exposure to a wide variety of microorganisms early in our childhood development—can lead to an overreactive immune system. The basic assumption throughout this chapter is that the natural environment matters deeply to the development of our allergic irritation. What our bodies are regularly exposed to—or not exposed to—has a significant and lasting effect on our immune functions.

Ultimately, the natural environment is only one more part of the complicated story behind the recent spike in allergies. By the end of this chapter, we'll begin to see why alterations to our lived, man-made environments—caused more directly by our modern lifestyles—are wreaking just as much havoc to our immune functions as changes to the natural environment. For now, however, let's explore how changing landscapes, changing technologies, and changing climates have contributed to the rise in hay fever and asthma rates in an allergy tale of three very different, yet all too similar, cities.

MANCHESTER, ENGLAND: THE INDUSTRIAL REVOLUTION AND THE HISTORY OF POLLEN

In the early eighteenth century, Manchester was a small, rural town nestled next to the rolling, green Pennine Hills in the north of England. With a population of fewer than ten thousand souls, it was an agricultural hamlet far away from the growing bustle of London in the south; its residents' lives reflected the pace and rhythms of the farmland and meadows that enclosed it. By 1819, the same year Dr. John Bostock gave the first description of hay fever, Manchester's population had grown to two hundred thousand. Only a few decades later, the population had doubled to more than four hundred thousand residents.

Along with this dramatic demographic explosion came an equally dramatic change to the city, its surrounding environment, and the way of life of its residents. The Industrial Revolution was in full throttle, and Manchester was at the heart of it. The burgeoning city—now England's second largest—had become one of the main centers for the production of cotton. Cotton mills, warehouses, and tenements began to dominate the landscape as the city's boundaries expanded ever farther. The neighboring farms, too, changed, as agricultural production raced to keep pace with the population explosion. Manchester, with its factories and farmlands, would be the backdrop to the discovery of one of the biggest environmental causes of allergies: pollen.

While it may seem obvious to us now, pollen was not a clear-cut environmental cause of hay fever in the early nineteenth century. The individual nature of the recently discovered disease—or how differently each patient would express symptoms—did not make it an easy task for physicians to discern any absolute causes.

Growing up in Manchester in the 1800s, Dr. Charles Harrison Blackley had witnessed all the social and environmental changes to the city firsthand. As people relocated from the countryside into the urban centers of England to search for work, their quality of life deteriorated. So, too, did their overall health.

Blackley had grown up suffering from bouts of summer catarrh, or hay fever, himself and was thus keenly aware of the early research and theories on the disorder, its causes, and treatments. By the time he began his investigations into the possible causes of hay fever in earnest in 1859, Blackley had suffered from the disorder for decades and was frustrated by the paltry understanding of the condition and the lack of effective treatments. There was also scant information on the etiology, or possible cause, of hay fever. Blackley's efforts to scientifically investigate the disorder, as he stated himself, were "personal."[1]

At the time, germ theory was beginning to gain ground as a serious scientific theory of disease causation. Blackley was curious if an external agent, or antigen, might be to blame for cases of hay fever. Because of its general mildness, and the lack of any known deaths caused by it, Blackley felt it was perfectly acceptable to experiment more methodically on the condition—initially on himself as a sufferer and only then, more slowly, on a few of his willing patients. Blackley took meticulous notes on exposures to different external agents, the time of day of exposure, and any resultant symptoms, determined to discover what might be triggering the attacks in the first place.

Of his own hay fever and asthma patients, the majority were either doctors or theologians. Blackley also noted the almost complete absence of the disorder among the farming class. He speculated that either the farming class didn't have the nervous predispositions that came along with education or repeated exposure to pollen on the farm caused them to become immune to the effects of pollen and other plant effluvia. Considering that more people were being educated in the mid- to late 1800s, the connection between education and hay fever seemed plausible. However, Blackley ultimately dismissed the emphasis that others sometimes placed on the nervous disposition of patients or their physical idiosyncrasies. He argued that since there had always been educated classes in England and hay fever had been largely unknown until the early 1820s, the true cause of the rise in hay fever must lie either in recent changes to agricultural practices or in the growth of cities. Although it was perfectly clear that some people had a predisposition to hay fever, he felt the important thing was to discover the condition's "primary exciting cause."

Farmland around Manchester, where Blackley resided, had expanded greatly. To accommodate the needs of a growing population, the type of crops being planted had also shifted. Instead of feeding cattle with vegetables and buckwheat, which had been the practice for decades, farmers had begun to give their animals hay as a staple food. The result was more hay production—and thus more effluvia in circulation throughout the haying season.

At the same time as agriculture practices and crops were shifting, the manufacture of cloth goods was moving into the cities. People who had toiled in small workshops or factories that were located in the countryside, near the fields, were relocating into the city to work in the newer, larger cotton mills. The factories also produced a need for more skilled workers who were educated. That education, Blackley mused, might be priming them toward hay fever, but he doubted it.

The urbanization of work meant that fewer people had regular, extended exposure to pollen in the fields, and that pollen itself was of a different type than it had been just a few decades prior. As the population of Manchester grew, so, too, had the need for more and more hay to feed more and more cattle to feed more and more people. Blackley surmised that this was the real culprit behind the soaring rates of hay fever he had witnessed in his own medical practice. To prove his thesis, Blackley began to methodically experiment with all the suspected causes of the time, including ozone, light and heat, odors of varying types, and pollens.

For his first experiments, Blackley filled a room with a substance (coumarin) that causes the odor of freshly mown hay, walked around the room at a quick pace so as to "vigorously" inhale the air, and noted the effects: absolutely none. He repeated the experiment with some of his patients to the same effect: No symptoms of hay fever appeared. When he repeated the experiment with odors of other plants such as *Chamomilla matricaria* (chamomile) and various fungi, the odors sometimes produced symptoms, including headache, but never the characteristic signs of hay fever or asthma. Then Blackley experimented with ozone. In the 1800s, ozone was thought to be a variation of oxygen produced by strong light hitting the leaves of plants. It was believed that ozone produced the strong odors associated with plants like juniper, lemon, and lavender. Ozone could be created using a combination of sulfur and potassium permanganate, and paper test strips could measure its presence in the air. Blackley performed many experiments where the ozone levels were shown to be high (per the strips) and never experienced any symptoms of hay fever.

Next up for experimentation: dust.

Dust, as Blackley's experiments showed, is particular to a time and place. Blackley argued that there is no such thing as "common dust," since its composition varies widely based on the geographic location, the house, the season, and even the time of day in which it is collected. Blackley noticed that dust could indeed produce some of the common symptoms of hay fever, such as sneezing and irritated eyes, especially during the seasons most associated with hay fever: May through August. In the book he wrote detailing his experiments, Blackley recounted walking down a country lane not often used, a few miles outside the city center. A vehicle passed him, kicking up a rather large cloud that enveloped him, and he was forced to inhale great quantities of dust. Immediately thereafter, he began a fit of sneezing that lasted for hours. His scientific curiosity piqued, he went back the very next day to kick up some dust himself to see if it would produce the same results. It did. Another attack of hay fever ensued. So he collected a sample from the road and took it back to his laboratory, where he inspected it under a microscope. There, looking at the dust on a glass slide, he observed copious amounts of grass pollen.[2]

Blackley felt he had discovered the smoking gun of hay fever causation: pollen. But he needed to conduct more experiments to be sure.

In his exhaustive experiments, published in 1873, Blackley reported on the physical effects of different types of grass pollen as well as the pollen of thirty-five other orders of plants. He varied the time of day and time of year, used fresh and dried pollen, sealed himself in rooms with pollen or walked about in air saturated with pollen. For each type of pollen, he would repeat the same steps. First, he applied it to the mucous membranes of the nose, to the conjunctivas of the eyes, and to the tongue, lips, and face. Then he inhaled it. Then, he introduced fresh pollen into small scratches on the skin of his upper and lower limbs, covering it with plaster (thus inventing the first skin scratch tests for allergies).

The results of all these experiments were largely successful. The pollen regularly produced symptoms of hay fever in varying intensity and duration. Blackley experimented with differing amounts and noticed that more pollen typically produced a stronger physiological response. When he experimented on patients, he adhered to strict control procedures—his patients never knew what they were being exposed to, so as not to affect the results. But mostly, he experimented on himself. His periodic experiments with pollen caused him stopped-up nostrils, violent fits of sneezing, headaches, bouts of asthma, and sleepless nights. Yet he continued his research program for several years.

Blackley found that temperature was correlated to pollen in the sense that below a certain temperature, plant growth was retarded and less pollen was produced. Different plants bloomed and produced pollen at different times under different environmental conditions. Whatever affected the pollen, Blackley opined, affected the allergy sufferer in equal measure. However, the size or shape of the pollen seemed to have little to no effect on the severity of symptoms. Neither did attempts to denature the pollen—as in boiling it—before applying it to the mucous membranes. That said, Blackley did observe that grains of pollen expanded when placed in water and surmised that part[3] of the problem during an attack of hay fever was that pollen grains also expanded upon contact with the moist mucous membranes lining the nose, throat, and lungs. By the end of this research, coupled with the fact that there were virtually no known cases of hay fever in the various British outposts and colonies in the tropics, Blackley had completely dismissed the then-popular notion that heat alone could cause hay fever.

Once he had ascertained that pollen was the direct cause of hay fever and bouts of asthma, Blackley began to examine his hypothesis that it was the quantity, and not the quality, of pollen that was of true importance to the hay fever patient. No one had ever attempted to measure the amount of pollen in the air or to categorize it by type or species. To test out his theory, Blackley began experimenting with several different homemade apparatuses.

After failing with several (rather ingenious) configurations, Blackley lit upon a simple design that produced consistent results. First, he painted glass slides with a 1-centimeter square of black varnish, to make it easier to see the pollen, and then coated them with a mixture containing glycerin. The glycerin was meant to mimic the sticky mucous lining of the lungs, providing a surface that the pollen would adhere to. Then he exposed the pieces of glass to the air.

Blackley used four glass slides facing north, south, east, and west to maximize the chances of an accurate count should the wind blow in multiple directions, carefully placing the slides about four feet above ground level to mimic the average height at which humans breathe in air.[4] The slides were set up in the middle of a grassy meadow used for haying, approximately four miles outside of Manchester.

After twenty-four hours, he took them back to his lab and viewed them under a microscope, meticulously counting the visible grains of pollen and typing them by species whenever possible.

Blackley repeated this experiment many times, varying the location of the slides.[5] The results were sometimes inconsistent, but Blackley posited that this was because he often found moths and butterflies clinging to the sides of his glass slides—presumably consuming the pollen grains. After several years of studies, Blackley found that the pollen count was always highest from May 30 to August 1. He also experimented with humidity and sunlight and found that the pollen count was higher in dry conditions, when the grass had been exposed to direct sun. Days of gentle rain followed by plentiful sunshine were the best conditions for the release of pollen.

All the evidence, it seemed to Blackley, was conclusive. Hay fever was clearly a physiological reaction to antigens in the immediate environment. That antigen was pollen, not heat, or ozone, or any of the other causes posited at the time. Despite his meticulous research being well received by such scientific luminaries as Charles Darwin, Blackley's findings would be ignored for several more years.[6] Due to the dominance of bacteriology and germ theory at the end of the nineteenth century, most physicians believed that hay fever and asthma were the result not of breathing in pollen, but of severe bacterial respiratory infections that had primed the lungs to be hypersensitive. The "bacteriology theory" of allergy causation—though inaccurate—would persist well into the 1890s.[7]

But by the time I was conducting research for this book, Blackley's ideas—especially his methods for taking pollen counts—had been more than vindicated, as we'll soon see.

CINCINNATI, OHIO: POLLENS AND PARTICULATES

It's the spring of 2019, and I'm sitting at a long, polished wooden table, looking at a large projection screen on the wall of the conference room at the Southwest Ohio Air Quality Agency. My host, Anna

Kelley, has been doing pollen counts and air pollution measurements in Cincinnati since 1984, when a Hamilton County commissioner who had allergies decided that they should start doing a daily pollen count for the city.[8] We're looking at a map of the greater Cincinnati area and Anna is explaining where the air quality monitors are located. A rotarod pollen sampler (a mechanical device that collects pollen grains from the circulating air) is located on the roof of the drab, seventies-style, concrete building that houses the agency, located near the geographic center of the city and close to Interstate 71, a major thruway that sees heavy traffic—and heavy pollution. I'm at the agency trying to piece together how we measure pollen in the air 150 years after Blackley first devised his own system. It's part of the puzzle of allergies—particularly hay fever and allergic asthma—and I want to better understand how the numbers we see on our weather apps and websites are generated.

When Anna asks me if I want to go up to the roof to see the pollen monitoring device, I immediately say yes. It's slightly unorthodox, she tells me, since technically only staff members are allowed up there, but an exception can be made.

"It's a short climb up a steep metal ladder," she says, "and then we'll need to step over a few structures. We'll be extra careful but do us both a favor: Don't fall."

It's a mild, partly cloudy spring day and when we emerge out of a rectangular hatch onto the roof, the wind immediately starts blowing Anna's gray bobbed hair and whipping at her stylish blue scarf.

The rotarod machine is much smaller than I had imagined it to be, a square white metal box sitting atop a black metal pole that is anchored to the roof by a large gray metal base. It looks a bit like a large square traffic light. On the underside of the top square sits the rotational arm that spins a plastic rod for one minute every ten minutes throughout the day. The rod is lightly coated in silicon grease on one side, snapped into the rotational arm, and swapped out again each morning. The collected rod is taken downstairs to the lab and stained, and then the pollen grains that have adhered to it are counted manually using a basic laboratory microscope.

The resulting numbers, when compared with seasonal daily averages, are used to determine whether or not pollen levels are high, medium, or low. Charles Blackley would feel quite at home doing this work more than a century after his death because it is incredibly similar to the method he originally devised. When I mention this to Anna, she laughs and nods, telling me that it's the same method she's been using for the past thirty years.

Anna and I watch as the rod spins. The machine is surprisingly loud as it whirs, so we are standing close together to talk. Anna explains that the reason the rod doesn't spin continuously is that the greased side would be so covered in pollen that it would be impossible to count the grains with the naked eye. There would simply be too many of them clumped together.

Directly behind the rotarod sampler stand seven large white metal monitoring devices for particulate matter. Each of them sits atop a long white metal pole and has a top that is continuously drawing the circulating air inside. The different monitors regularly test for common air pollutants, including ozone, carbon monoxide, sulfur dioxide, and nitrogen oxides. These machines, unlike the noisy pollen monitor, are almost silent. Each one is attached to a cord that snakes back down into the building, to the monitoring room, where stacks of equipment provide real-time readouts of air quality. We climb back down from the roof to visit that room, which is extremely loud and quite warm. Anna tells me that each machine is incredibly expensive to purchase and maintain. They have specialized staff whose job it is to regularly check and recalibrate each monitor.

The Environmental Protection Agency (EPA) sets standards that each local agency in the United States has to meet for measuring air pollutants. In other words, air quality is highly regulated and fully funded. There is no such standard or national coordination, however, for pollen levels, so pollen counts are an entirely homegrown affair. Each local agency keeps its own data and decides—based on decades of collection—what are "high" or "low" counts for its own area. What's more, Anna explains, the rotarod is not considered the "gold standard" for pollen measurement.

"That's the Burkard monitor," Anna says. "It's a big plate that's greased, and air is drawn in over time. The pollen and mold spores are deposited through these puffs of air."

The Burkard monitors show a better sensitivity for mold, grasses, and weeds than rotarod samplers (both machines work equally well for tracking pollen under normal conditions). However, they are also far more expensive. And since pollen counting isn't federally mandated, most air quality monitoring stations have to fund pollen monitoring with local funds. Anna tells me that the rotarod data is reliable enough (the agency's counts are certified by the American Academy of Allergy, Asthma & Immunology), and that they recertify the machine annually to ensure it's working properly. In other words, the rotarod is staying on the roof—for now at least.

Back in the agency's lab, Anna shows me a rod itself under the microscope. As I squint and peer into the eyepiece, my eye focuses on dozens of small, mostly circular, pink-stained objects. Anna asks me to try typing one by describing what a typical oak tree pollen grain should look like. To the untrained eye, all of the grains look too similar to distinguish from one another. I laugh and give up after only a few moments. She tells me that it takes a long time to get good and efficient at this work.

Each morning, Anna or another employee will sit down at a small metal stool in front of the microscope and tally up the individual pollen grains on the cylinder by hand, using a micrograph book as a reference. The micrograph book is a collection of digital photographs taken through a microscope of the different species of pollen in the area.

It takes Anna, who has years of experience, two or three hours each morning to complete the count. Workers have been doing it this way for decades. Sometimes the stains used on the pollen can make the grains difficult to identify—maple will look like oak, depending on the way it attached to the rod. It takes time to learn the different seasons for each tree and grass and to begin to guess more accurately which pollen belongs to which plant. It can be an even more arduous task during particularly heavy pollen seasons because everything is blooming at once and the rod is covered in pollen. The other main challenge, Anna explains, is the introduction of non-native species into the area. Anna tells me that a lot of people are planting Chinese elm now, but it pollinates in the fall. Native elms pollinate in the spring. This extends the pollen season for "elm trees" as a grouping.

"We do the best we can do," Anna says.

Once the manual pollen counts are done, they're uploaded to the Web. The daily pollen reports that you see online or in newspapers are yesterday's counts, not today's. There is no "real-time" data on pollen; it's always lagging a day behind. That's okay, however, since pollen levels rise and decline in increments—unless there's rain, of course, and then the rates temporarily plummet. The numbers are local, too, meaning that if you see a "high" count for elm, it is "high" by Cincinnati standards. It's not a nationwide average. Each city's pollen count will be different and will have a different threshold for what counts as "high" or "low" pollen loads. (The exception is reporting stations for the American Academy of Allergy, Asthma & Immunology, which use standardized levels nationwide.)

Anna tells me that, in addition to the local average levels, she also tries to put out the agency's daily raw counts, in an effort to aid allergy sufferers trying to decipher which types of pollen might be causing their hay fever or asthma symptoms. Here, in Cincinnati, Anna tells me that she hasn't noticed a dramatic change in the length of the pollen season or in the daily amount of pollen circulating in the air. But she has noticed that her own respiratory allergies to grass and weeds have gotten slightly worse over the last five years or so, from "nothing" when she first began counting pollen to moderate. During that same time frame, Anna has seen new pollens enter the region, as landscapers have introduced one or two new species to the area that are non-native. But other than that, she hasn't seen any of the more dramatic changes that other monitoring stations—typically those located in cities at much higher and lower latitudes—have over the last decade.

For now, Cincinnati begins measuring pollen in February, when the cedar trees begin to pollinate, and continues daily counts until

Thanksgiving in November—the very end of the local pollen season. In places like Alaska, Minnesota, Wisconsin, Louisiana, and Texas, the vagaries of climate change can significantly alter the length of the pollen and mold seasons, making life more miserable for hay fever and asthma sufferers in those regions.

Unlike pollen counts, air quality sampling takes place year-round, rain or shine. Many people call the Cincinnati agency confused about the meaning of the label "air quality," especially during the height of pollen season. They often, and understandably so, conflate pollen/mold counts with air quality levels, despite the fact that those two measurements are not at all the same. Daily air quality maps and indexes, Anna emphasizes, represent the number of pollutants and the amount of particulate matter in the air—not pollen or mold.

"They'll see a high pollen count and call up to complain that the air quality says moderate for that day," Anna says.

Air quality data for the city of Cincinnati is collected not only from sampling instruments located on the agency's rooftop but also from instruments scattered at different locations throughout the city to measure particulate matter of 2.5 microns or larger. There are strict federal requirements about where the monitors are installed (some monitors need to be placed within fifty yards of a road) and how the air is monitored (which machines are used and how they are calibrated), as well as different probe heights for different pollutants. All the resultant data is fed back into the Ohio EPA district office.

Measurements of pollutants are standardized, and the air quality indexes we see on our weather apps or on the news—from green (good), yellow (moderate), orange (unhealthy for sensitive groups), to red (unhealthy)—are regulated by the EPA. Unlike the pollen levels, pollutant levels do not vary from place to place. Data on air quality is available hourly, so people can log on to the agency website and see real-time data.

"People really don't understand the difference between pollen and particulates," Anna tells me. "We only measure 2.5 microns, but there are particulates—called ultrafine—that are smaller than that.

Those are dangerous, too, but the 2.5 have known health effects. Over time they build up. But we don't think about those things."

Like pollen, particulate matter can also negatively affect our ability to breathe, but it is a different, if intimately related, problem. "Particulate matter" is the term used to describe any microscopic substances—liquid or solid—that are suspended in the air. The ones we measure can be sorted into two bins: those that are labeled as "coarse" particles, which are 10 microns in diameter, and those that are labeled as "fine" particles, which have a size of 2.5 microns. "Ultrafine" particles (UFCs)—or those with a diameter of 1 micron—are not typically directly measured at all. (Important sidenote: UFCs can be measured by using condensation particle counters. The likeliest reason they are not measured is not because we can't, or even that the cost of the counters themselves is prohibitive, but because there are no governmental mandates to measure them. In other words, we don't want to as a social and political collective.)

How small are all these invisible pieces of matter? A single micron is equivalent to 0.00004 inch. A red blood cell is 5 microns in diameter. A single strand of hair is about 75 microns. For context, pollen grains are anywhere from 10 to 1,000 microns, depending on the species. In other words, particles that are 2.5 microns or smaller are incredibly, almost unimaginably, tiny.

And yet, ultrafine particles are everywhere. They are produced by diesel combustion engines, factory smokestacks, and coal-burning energy plants. They get into the air from smoking cigarettes or burning wood—even from the normal daily activity of cooking in your kitchen. Out of all the facts and statistics I encountered during my time researching this book, this one might haunt me the most.

Everywhere I go now, I think about what might be in the air I'm breathing. Millions of children and adults around the globe are exposed daily to high levels of this type of air pollution, with its ultrafine particles. For those of us who live in major cities, or even near them, it's largely impossible to escape. Once you know more about these particles, the idea of "getting some fresh air" becomes almost tragically laughable.

I ask Anna if doing this job has made her more aware of the air that goes into her lungs. When she breathes, does she think more about all the invisible things—pollen, mold, ozone, particulates—that she's inhaling into her body?

"I've been in this so long," Anna reflects, "it's second nature. But, yes, I am more aware. I love a campfire. A fire on a cold night in the fireplace? I love it. But I also recognize the particulate matter coming off of it."

Then we have a short conversation about all the summer and fall wildfires in the western half of the United States, and their likely effects on short- and long-term respiratory health. Although particulate matter greater than 5 microns is visible to the naked eye (think of the pictures of hazy air in Los Angeles or Beijing), most of the more nefarious pollution in the air is utterly invisible. Because people can't see particles that are 2.5 microns, Anna laments, they don't think about it. At all. On clear days, when the air quality index is high, and thus air quality is poor, people almost can't believe the science behind it. She smiles a little wryly and says, "They just see a beautiful day."

Air quality has been monitored in Cincinnati since 1976, and the Southwest Ohio Air Quality Agency has slowly become a repository of historical data on local air pollution.[9] The agency has set up monitoring stations near major roads and highways, in industrial areas, next to a steel production facility, and near former and newer coke plants. (Coking is a heating process whereby raw coal is processed into a high-carbon material called "coke.") Air quality monitors are usually placed in areas with higher populations or where emissions might be higher. Researchers in the area, especially at the renowned Cincinnati Children's Hospital, often use the agency's data set to look for correlations between air pollution and a wide variety of health conditions—including allergies and asthma.

In the late 1990s, researchers showed that air pollution can act as a transport mechanism for allergens. Diesel exhaust, so ubiquitous in urban areas, is one of the main culprits. As Dr. Patrick Ryan, an environmental epidemiologist at Cincinnati Children's Hospital, explains, "It can bind pollen to the surface of the particles and actually serve as a way that those pollens get deposited deep into the airways. We were interested in diesel exhaust particles initially because we thought these particles could drive an immune type of reaction and then lead to allergies in kids and subsequently to asthma, which was an open question at the time: Does air pollution cause asthma or just exacerbate it?" In other words, since diesel particles are so small, after the pollen binds to them, both can be inhaled much deeper into the lungs than pollen could be on its own, making them more likely to ignite an immune response.

The Cincinnati Childhood Allergy and Air Pollution Study (CCAAPS), first begun by Ryan's former adviser and fellow environmental epidemiologist Dr. Grace LeMasters, shows that people who live in areas with greater exposure to air pollution, especially diesel exhaust, have a higher risk of developing respiratory allergies and asthma. From October 2001 to July 2003, the CCAAPS recruited 762 infants from the greater Cincinnati and northern Kentucky area with varying exposures of air pollution, based on their parents' addresses on birth records. Ryan, a graduate student at the time, remembers that the study was specifically interested in recruiting babies who were living in households adjacent to major roadways (with one thousand trucks or more passing by their homes each day). The CCAAPS consisted of two distinct birth cohort groups: one that lived less than four hundred meters from a major road, and one that lived more than fifteen hundred meters away. All the children enrolled in the study were then regularly tracked for the telltale signs and symptoms of respiratory disease, beginning at age one, then again at ages two, three, four, seven, and twelve. (As of this writing, the original birth cohort members are twenty to twenty-one years old.) The CCAAPS is unique in that it was only one of twelve birth cohorts being studied at regular intervals over an extended amount of time in an attempt to capture the ill effects of living close to sources of daily air pollution (some of the other cohorts were located in urban areas in Michigan, Massachusetts, Arizona, Wisconsin, and New York).

"Basically, what we saw," Ryan says, trying to sum up a mountain of data for me, "is that the kids who were exposed to the highest level of air pollution early in their life were more likely to develop asthma by age seven. We have kids that had high exposures early in life and then moved to areas with lower exposures. What we saw is that if they had early and sustained exposures, even if they moved to better areas, they were more likely to go on and develop asthma. We also saw that the onset of their symptoms was earlier than [in] those that didn't have the same level of exposures."

When I press him, Ryan is still hesitant to say that environmental exposure alone—in this case diesel particulates in the air—is the true culprit that has caused significantly higher rates of allergies and asthma. He explains that "a kid that lives right next to Interstate 75 [one of the main traffic arteries in the Cincinnati area] also probably has poor access to food, poor access to healthcare, worse indoor environmental exposures, more mold exposures, and cockroaches. So to point to air pollution as the sole cause is just impossible. But, at the same time, I'm also absolutely certain it's a contributor." And who is more likely to live near these busy highways? Families living near or below the poverty line. Our most vulnerable citizens economically are often also our most vulnerable biologically to exposure to greater levels of pollutants.

That data, Ryan stresses, is crystal clear on this point: As air pollution rates drop, asthma rates also drop. This fact seems relatively straightforward and matches our base intuition about the relationship between healthy air and healthy lungs, but the study adds critical scientific evidence to Charles Blackley's two-hundred-year-old conjecture that industrialization and modern technology have direct negative effects on our respiratory health. Blackley, I'm certain, would not be at all surprised by the outcome of the CCAAPS. Perhaps more interestingly and surprisingly for all of us, however, is a related finding that those same levels of exposure to particulate matter are linked to changes in our brains that increase our risk of anxiety, depression, and perhaps even dementia.

Ryan explains that diesel particulates are so small they can seep through our blood vessels and nose cavity directly into our brains, where, his latest research suggests, they might be able to alter neural pathways. In fact, his team has found that greater exposure to diesel exhaust is associated with higher levels of anxiety and depression in children by the age of twelve.

In Cincinnati, a busy bridge that spans the Ohio River might have as many as seventy thousand diesel-engine trucks crossing it per day. People living in the hilly areas of the city get better air than those in the valley because, depending on the weather and the season, air often gets trapped in the river valley. Ryan's team of researchers is now busy trying to understand the finer details of daily exposure by getting study participants to use personal, wearable air monitors. Ryan can estimate the level of exposures outside in one area by using Anna's data. But regular people don't stay outside and don't stay in one location throughout the day, so even a home air quality monitor isn't going to tell him what he really needs to know: What is the real-time level of daily air pollution that any one individual encounters, and how is that related to their overall health?

Allergies and asthma are just two of the possible ill effects born of the combination of pollen and particulates. As we've seen before, those of us who suffer with allergies are simply the forerunners of what is likely to be happening to all of us: the deterioration of our lung health. And before you rush out to purchase an air purifier or other filtering device, know this: They probably don't help, and they might actually make things *worse*.

A study published by the American Thoracic Society showed that exposure to filtered air plus an allergen produced worse symptoms of respiratory allergy than either the allergen alone, nitrogen dioxide alone (NO_2 is formed during the process of burning fossil fuels and thus is found in high quantities in diesel exhaust), or unfiltered air plus the allergen.[10] The air had measurably higher levels of NO_2 *after* HEPA filtering. What this study ultimately suggests is that there is no easy technological solution to the problem of pollen, air pollution, and allergies. But perhaps knowing this will make us—as a species— more eager to work together to curtail the production of particulate matter in the first place.

CHANDIGARH, INDIA: PARTICULATES, POLLEN, AND FUNGAL SPORES

Since particulates can carry pollen and make their impact worse, it's nearly impossible to talk about asthma—or any respiratory issue for that matter—without talking about two things conterminously: the first, pollution and its effects on lung function, especially in young children, and second, the effects of climate change on the average levels of pollen and spores that are circulating in the air. One of the arguments for the rise of allergies in the twenty-first century is the aforementioned hygiene hypothesis (which we'll go into in more depth in chapter 6), or the idea that as we moved from living in rural farming communities into cities, and as family size started to shrink, we were exposed to fewer "good" bacteria during our developing years. The thinking was—and in some quarters, still is—that without an adequate supply of exposures to mild microbes, our immune systems do not become well trained to recognize friend from foe. From this vantage point, our immune cells are rather like unruly children who are bored by our hyper-clean environments; they come up with things to do to entertain themselves and it's not always to our benefit.

In the past decade, the rising rates of allergies in places where people are assumed to have a more diverse microbiome in their environments has directly challenged this theory. Allergy rates in so-called westernized (read wealthier) nations are frequently compared with those of non-westernized (read poorer) nations. Although rates of all allergies remain higher in westernized nations, the rates are quickly climbing in other places. It turns out that "the environment" that seems to produce more allergies isn't so simple to describe—or to avoid.

It's a cold, winter evening near midnight, and I am on a Zoom call with Dr. Meenu Singh at her home in Chandigarh, India. She's sitting outside on her patio, shielded from the morning sun. She pans the camera for me, showing me the lush greenery around her. Tall trees, vibrant grass, bushes of varying shapes and sizes, some in bloom. Encased as I am in a blanket from the waist down, I am jealous. As we chat, birds are calling to one another in the background.

Singh is a professor of pediatrics and the head of the Pediatric Pulmonology, Asthma and Allergy Clinics at the Advanced Pediatrics Center at the Postgraduate Institute of Medical Education and Research in Chandigarh. She's been seeing asthma and respiratory allergy patients in the clinic for decades. When I ask her if allergies are a growing problem in India, she nods.

"We never used to see cases of eczema and now it is quite prevalent," she says. Food allergy has also recently appeared in India, which heretofore was very, very rare. "Peanut allergy was never heard of in India," Singh explains. "But now we're getting cases. I don't know whether it is the hygiene hypothesis alone; there could be other factors also."

Singh tells me that her specialty, asthma, had seen big increases over the past few decades, but its prevalence in children in India has recently stabilized to around 3–4 percent (still a massive rise from its reported rate of 0.2 percent in the late 1960s). Asthma's cultural and social profile has shifted over time. Once a disease associated with elite urban intellectuals (much like hay fever), asthma now primarily has a reputation as a disease of poor urban dwellers.[11] In fact, the World Health Organization (WHO) reports that there are more deaths from asthma in low- and lower-middle-income countries, likely due to a lack of medications to control the condition in more severe cases and less access generally to medical resources, making even diagnosis of the condition more challenging. In 2019, the WHO estimated that 455,000 people died globally due to complications from asthma.[12] Those most affected by asthma complications are not only the poor, but young children and the elderly, with the elderly more at risk for death.

The WHO defines asthma as "a chronic disease of the air passages of the lungs which inflames and narrows them," characterized by "attacks of breathlessness and wheezing, which vary in severity and frequency from person to person." As the historian Mark Jackson notes in his history of asthma, and as we've already discussed in our examination of how allergy is diagnosed and tracked, asthma is not a "one size fits all" lung disorder. In fact, it's more likely to describe a consistent pattern of symptoms than a single cause. That's because the fundamental causes of asthma are not wholly understood. We know the risk factors associated with the development of asthma: frequent childhood infections, exposure to cigarette smoke or high levels of indoor and outdoor air pollution, and a genetic susceptibility. Yet as of this writing, there still isn't a single, definitive definition of asthma. Its association with respiratory allergy is also still debated, yet more and more practitioners like Singh make little distinction between allergic asthma and other forms, like exercise-induced asthma, since most asthma patients also have allergies and are triggered by a variety of environmental irritants. When I ask Singh if she's seen more respiratory allergies, or if her asthma patients' symptoms are being triggered more frequently or worsening, she nods her head.

"Even in Chandigarh, which is a relatively clean city, we do have quite a bit of particulate matter. And people who are staying close to the crossings and the main roads, they have higher prevalence of allergies."

This echoes what I heard from Patrick Ryan in Cincinnati. Exposure to particulate matter—and poverty—exacerbates allergy and asthma everywhere, no matter what other exposures patients have. In fact, the WHO's prevention focus for asthma is on lowering levels of pollutants, not allergens, because reams of epidemiological evidence show that continuous exposure to NO_2 leads to increased rates of childhood asthma. And while allergens might trigger or worsen asthma attacks, it is unclear if they play a role in causation. Ryan suspects that the combination of pollen and particulate matter is the one-two punch in childhood development that leads to allergic asthma. And children living in poor urban areas are more likely to have higher concentrations of particulate matter in the air they breathe. In this way, Chandigarh is no different from Cincinnati.

Most of Singh's patients are not wealthy. Since her clinic is a government-run facility, her patients tend to be those who cannot afford to go elsewhere for their treatment. Sometimes the wait to be seen at the allergy and asthma center can be several months long. And yet, Singh tells me that there is no end to the waitlist; the problem, from her vantage point at least, is that the demand for allergy care seems to be growing ever larger. There is simply not enough staff to keep up with the need.

When I ask Singh what she thinks is driving the increase in allergies in her area, she points to the dual problem of changes in the local environment and Indian lifestyles. It's the same story, really, as everywhere else—only the details change. Farther south, she explains, in Delhi or Mumbai or Chennai, everything is concrete and crowded, and there is more pollution and early exposure to cigarette smoke. In Chandigarh, there is also air pollution, but also more vegetation—and thus more pollen in circulation. That's due in part to the way the city was designed.

"This city came out of scratch," Singh explains. "They built this city after Indian independence and it was designed by Le Corbusier, so they planted a lot of trees."

Chandigarh was built using a "garden city" model, which was developed in England as a response to rampant industrialization. At the turn of the twentieth century, an English town planner, Ebenezer Howard, wanted to design an ideal, utopian city that would marry the best of urban life with the best of rural agricultural life. The so-called garden city would contain more green spaces—and thus more plant life—to counteract what was viewed as the ugliness of modern factories and cramped, makeshift housing. In Chandigarh, the elite ruling classes would live in homes surrounded by lush greenery. The streets and boulevards, too, would be lined with trees. Artificial hills were designed. The total effect was a "greener" city.

The plan made Chandigarh a beautiful city, but one that has a huge pollen season. Singh laments that unlike many other cities, Chandigarh doesn't currently do daily pollen counts. That information, she argues, is desperately needed. With pollen tracking, she might be able to provide better treatments for her respiratory allergy and asthma patients.

It's not just diesel engines and factory exhausts contributing to the woes of Indian asthma patients. The other problem with local air quality in many areas of India is crop burning, Singh suggests: "What they do, when they cut the crops, is . . . whatever is left behind . . . they burn it. And that leads to a lot of smoke that worsens the air quality. We need to learn how to manipulate our environment in these situations." But Singh also points out that Indians will have to rethink their lifestyle choices in order to curb rising asthma rates. "It will probably start with education about restricting vehicle use or working from home."

In a way, the COVID-19 pandemic was a boon for Singh's allergy and asthma patients because it was a boon for air quality in India. The air temporarily cleared a bit; it had less particulate matter. That, combined with the wearing of masks throughout the pandemic, likely led to Singh's patients having fewer asthma attacks. Her respiratory allergy patients, too, had an easier time coping with the heavy pollen season in April and May of 2020. As the pandemic starts to recede and public health measures are slowly being lifted, she's thinking of designing a scientific study to measure the effects of mask wearing on the rate of allergy attacks. The downside to the pandemic for her patients, of course, is that those with indoor allergies to things like mold or mites have worsened. There is no winning with allergies, it seems, even when the outside air temporarily clears.

In fact, one of the most common allergies that Singh sees is fungal allergies. About 20 percent of her patients with poorly controlled asthma will go on to develop a severe lung condition known as allergic bronchopulmonary aspergillosis (ABPA). While rare, ABPA is more likely to develop in patients with poorly controlled asthma who become sensitized to multiple species of the *Aspergillus* fungal family (there are 837 species in total found around the globe).

"This is a very bad disease," Singh says, "because it makes asthma so much worse. It destroys the lungs, which otherwise doesn't happen in simple asthma."

An overgrowth of fungus is to blame. Singh points her finger at increased construction, poor agricultural practices, and a gradually changing climate for the increase in fungal spores in and around Chandigarh. Fungus thrives under wetter, warmer conditions, and a lot of the new construction sites in the city are located on former crop fields that have high water tables. "All these newer houses have a dampness problem," Singh says. In the north of India, where Chandigarh is located, there are not many dust mites to worry about—a blessing for her patients. But in the much warmer south, Singh explains, patients are exposed to both fungus and increased dust mites—a ruinous combination for asthma and ABPA. And unfortunately, there are no good treatments for ABPA.

"There are no uniform guidelines. People use steroids; people use antifungal agents. But we have to continue to treat the asthma. Their IgE levels are sky-high."

Because of climate change, and the wetter and warmer weather conditions it is producing across the globe, fungus sensitization is a rising problem, especially in South Asian countries. Those fluctuations in temperature can—and do—produce changes in when and how plants and fungi bloom, wreaking havoc for an increasing number of respiratory and allergy patients worldwide.[13] And although Singh has hope that the future might bring better, more effective, and cheaper treatments, she has no illusions about the path we're on. Her city may be beautiful, but it is also filled with particulates, pollen, and an increasing number of fungal spores. The air quality, despite the pandemic, is unlikely to get better in the next decade. She just hopes that her clinic will be able to keep up.

OUR IMMUNE SYSTEMS AND OUR CHANGING NATURAL ENVIRONMENT

Allergy is beyond question the most important biological and medical problem that exists or ever has existed, for it represents the pathology of the reaction of man and the lower animals to their environment—to the air they breathe, the physical agents such as light, heat, and cold, to which they are exposed, the food they eat, and the various parasitic organisms which may invade them.

—H. W. Barber and G. H. Oriel, "A Clinical and Biochemical Study of Allergy," *Lancet*, November 17, 1928

What Manchester's past, Cincinnati's present, and Chandigarh's possible future tell us is that our natural surroundings—in all of these cases, specifically the air we breathe—can have a dramatic effect on our immune function. Our allergy risk is not just genetic, or partially inherited, but environmental, or triggered by the invisible particles we encounter on a daily basis. Perhaps some of the strongest evidence in support of the idea that our environments are to a certain extent causing the increase in all our allergies comes from studying our white blood cells themselves.

New research (2020) from the Wellcome Sanger Institute, a leading nonprofit research institute in the United Kingdom, shows that our T cells are not simply "switches" that turn on or off in response to exposure to an antigen. Rather, the more "experienced" the T cells are, or how much prior "training" they have had in recognizing and responding to a particular type of antigen (say, dust mites), the more diverse their responses can be—forming a spectrum of immune system responses. In the Wellcome study, the more the T cells had reacted to a signal in the past, the faster they reacted in the present, no matter which response they had chosen.

"Previously people thought that memory T cells had two stages of development," writes Dr. Eddie Cano-Gamez, one of the study's lead authors, "but we discovered there is a whole spectrum of memory experience."

Researchers found that when "naïve," or inexperienced, T cells were given a specific chemical signal, they first responded to it by calming down or limiting the immune response. But with more "experienced" T cells, or cells that had encountered the antigen before, the reaction was the total opposite. Those more experienced immune cells ramped up inflammation. In other words, the more times you're exposed to cedar pollen and particulate matter, the worse your reaction to them might be. In places with heavy pollen loads and poor air quality, that means more respiratory allergies, more asthma, and perhaps more severe symptoms.

The fact that so many asthma patients also struggle with respiratory allergies doesn't surprise researchers. Dr. Robert Schleimer, former chief of the Division of Allergy and Immunology and now professor of medicine (allergy and immunology), microbiology-immunology, and otolaryngology in the Department of Medicine at Northwestern's Feinberg School of Medicine, explains that an asthma patient has a 90 percent probability of having hay fever. Dr. Schleimer uses the metaphor of fans doing the wave in a giant football stadium to describe the lining of the nose and its mucus. Particulates and pollen that enter the nasal cavity are moved very quickly out of the nose by cilia (microscopic vibrating hairlike structures on the surface of some cells) in a wavelike motion. Mucus containing those particles then drops down the back of the throat, where it makes its way into our stomachs.

"We swallow about a liter of mucus a day," Schleimer says, shrugging his shoulders. And that's on a normal day, one without any signs of an allergy or an infection. "Studies have shown that if you put a paper with saccharin or sugar into the nose, people will taste the sweetness about twenty minutes later—the approximate time it takes the nose cilia and mucus to clear it."

The mandibular glands, sometimes called Waldeyer's ring, are located around the area in the throat that your nose filters into. These glands are part of the body's lymphatic system and include your tonsils and adenoids. Their job, so to speak, is to sift through the mucus and decide if it contains anything dangerous or harmful. If so, then they can signal an immune response, which will extend to the lungs.

"So the process I've just described is part of what is called the unified airway hypothesis," Schleimer explains. "The unified airway hypothesis argues that allergic inflammation, when it occurs in the respiratory tract, tends to occur everywhere in the entire respiratory tract."

The theory is also supported by more than two hundred years of observations and scientific research on hay fever, asthma, and the relationship between exposure to antigens in the air and the development of respiratory allergy. And because of climate change, agricultural growing seasons—especially in the North—are elongating. EPA maps from 1995 to 2015 show an average increase of twenty-one days in the pollen season in Minnesota, fifteen days in Ohio, and six days in Arkansas. A study conducted at the University of Maryland between 2002 and 2013 and involving three hundred thousand respondents showed that hay fever increased whenever the timing of spring changed.[14] The prevalence of hay fever increased by as much as 14 percent when spring came early. That's bad news for the millions of young children in cities like Cincinnati and Chandigarh who are being exposed to consistently high levels of particulate matter, pollen, and fungal spores.

Take ragweed—one of the biggest natural environmental triggers of respiratory illness. Ragweed is a flowering plant native to the Americas. It is notorious both for its proliferation and for its pollen. In many ways, the story of ragweed in the past two hundred years has become the paradigmatic example of how environmental changes can have an enormous impact on allergies. Ragweed is very sensitive to any changes in the level of carbon dioxide, or CO_2. Its production of pollen intensifies with higher CO_2. Rising levels of CO_2 in our atmosphere, while terrific for ragweed, are disastrous for allergy sufferers everywhere.

But the problem doesn't stop with ragweed.

Dr. Richard Primack, a biology professor at Boston University, knows a lot about pollen—both personally and professionally. When he was a graduate student studying the ribwort plantain, a flowering species of plant that grows best in environmentally disturbed landscapes, he developed a severe respiratory allergy to its pollen. Even after he had stopped working with the plant, the allergy lingered for years. It's one of the occupational hazards, he tells me, of being a botanist. Nearly every botanist becomes allergic to something at some point in their careers. Since they are encountering the same pollen at greater levels than they would outside the research laboratory, and more frequently, their immune systems have more opportunities— and incentive—to respond negatively.

When I call him, it's mid-fall and seventy degrees and he's eager to talk about the production cycle of natural allergens. The topic is right up his research alley. Dr. Primack's biology lab at BU focuses on the effects climate change has on the timing of biological events like spring pollination. When I ask him about the proliferation of pollen and mold spores, he's more than happy to discuss the many changes he's observed over the past four decades.

In sum, if you think your seasonal respiratory allergies have been getting worse each year, you're probably right. Pollen and mold spore levels have indeed been shifting. Several climate factors are currently intersecting to compound the problem.

Most obviously, the temperatures are warming. Spring seasons are, on average, happening much sooner—beginning as early as February in some locations—so plants and trees that respond to warmer temperatures are flowering earlier, too. At the other end of the growing season, fall temperatures are much milder, which allows plants to flower for extended amounts of time.

"In the New England region, where I'm from, there would generally be cold weather starting in late September and a killing frost sometime in early October," Dr. Primack explains. "And that would really stop all the grass plants, the ragweed, and other pollen-producing plants from flowering. And what's happening now with climate change is that the weather is fairly warm throughout September and into October. This year, we had warm conditions and plenty of rainfall in October, so plants like ragweed just kept growing and producing more flowers."

When plants like ragweed emit pollen long into the fall months, it means extended misery for those of us who suffer from ragweed allergy. But climate change is not just creating a bigger problem for respiratory allergy sufferers. What other allergenic plant loves the new weather patterns? Poison ivy.

"Poison ivy is just dramatically more common now than it was when I was growing up," Dr. Primack says matter-of-factly. "These types of plants are spreading, they're more prolific, and they're in places where they didn't occur before."

Some plants benefit from air pollution itself. More circulating carbon dioxide is advantageous for plants like ragweed and poison ivy. But plants also really love higher nitrogen levels.

"In the past, soil nitrogen was a limiting nutrient for many plants,"

Dr. Primack says. "But because of increased burning of fossil fuels—like oil, coal, and natural gas—there's more nitrogen dust being generated. And this dust, when it falls on the ground, fertilizes the soil. So plants like ragweed are able to take advantage of higher nitrogen in the soil, greater amounts of CO_2 in the air, and warmer temperatures to grow more prolifically than they did in the past and to produce more pollen."

A bevy of environmental changes have also produced better breeding grounds for more invasive species of plants. Places like Southern California, Arizona, and New Mexico are seeing increased levels of pollen due to an influx of invasive species of grasses. The Midwest is seeing the effects of milder weather with grasses flowering much later than they typically do. And in the South, with its already high levels of humidity, things are getting wetter and hotter—a bad combination for those of us with mold allergies. Those are ideal conditions for mold growth and there are more mold spores in the air.

In essence, then, there isn't a single area within the United States that isn't seeing the direct effects of climate change on allergens. And while we may all be dealing with slightly different problems—from mold to ragweed to oak trees to invasive grasses to poison ivy—they're all pointing us toward greater irritation.

Pollen levels, in general, are expected to double by 2040, and the pollen will be more "potent" (its peptide levels will rise, likely worsening our immune system reactions). A recent study suggested that longer pollen seasons will lead to more emergency room visits for allergic asthma.[15] The study focused on oak pollen, which already sends approximately twenty thousand people to the ER each year in the United States alone. Research done at the Mayo Clinic in 2017 linked climate change to an increase in CO_2 levels that led to an increase in fungal growth.[16] The study found that exposure to fungus lowers cell barriers, causing cell inflammation that can worsen allergies. Climate change is also causing worse flooding and higher temperatures around the world. And that means more mold, as we've already begun to witness in places like Chandigarh and in New Orleans, where allergy rates began to skyrocket after Hurricane Katrina. Climate change is also altering weather patterns, and storms exacerbate respiratory allergy and asthma symptoms—a phenomenon called thunderstorm asthma. Rainfall ruptures bioaerosols and lightning strikes fragment pollen grains, then increased winds distribute those ruptured fragments for miles and miles. In 2016, a thunderstorm asthma event in Melbourne, Australia, sent more than ten thousand people to the emergency room with difficulty breathing over just two days.

All of this gives credence to the argument that changes to our natural environment have been, are, and will continue to affect our immune system function, causing worsening allergies. But even if we buy into the unified airway hypothesis, and accept all the scientific evidence supporting it, then what caused the equally dramatic rise of eczema, or skin allergies, and food allergies? Is our natural environment to blame for them as well?

As Dr. Elia Tait Wojno, an immunologist at the University of Washington, said to me, "It's complicated."

Tait Wojno studies allergies in dogs, arguably our favorite companion species. Our cats, dogs, and birds are special in that they live in our homes, share all our spaces, and eat food that we've manufactured. Tait Wojno points to the fact that our pets and farm animals also experience allergies to support the idea that it's our total environment that is causing the bulk of the problem. It's not just our immune systems that have gone wonky but theirs, too.

"I think there is an argument there that there is something going on environmentally," Tait Wojno said, "whether that's food, whether that's industrialization, chemicals, toxin, just all of the above, in an evil combination."

It's the missing part of that "evil combination" that we'll turn to next, as we explore all the various changes we've made to our lifestyles that may be negatively affecting our—and our pets'—immune function.

CHAPTER 6

Are We Doing This to Ourselves? The Modern Lifestyle and Allergy

Elizabeth is an engineer in her late thirties with three wonderful children—all of whom happen to have had some form of allergy. Her older daughter, Viola, age twelve, had eczema as a baby and currently has environmental allergies to pollen and food allergies to corn, tree nuts, and peanuts. Elizabeth's son, Brian, age three, also had eczema as a baby. He's subsequently developed food allergies to peanuts and barley—at least, those are his known food allergens. Elizabeth fears there could be more. Her younger daughter, Amelia, age five, had a dairy allergy as an infant, but is now just lactose intolerant. In terms of allergy, she's the easiest of the three children.

By the time I hear her story, Elizabeth is already an old hand at dealing with her children's irritated immune systems. Her own understanding of allergy is largely based on her experiences over the years. She's also part of a support group that she began for parents of children with corn allergy and is heavily involved in trying to educate other parents about food allergy.

Elizabeth says that she and the other moms in her group have developed "theories" over the years about why their children have allergies. Her own is that Viola and Brian both went to the emergency room with high fevers as small babies, and both were given antibiotics as a precaution. Elizabeth blames the antibiotics for altering her children's gut microbiome, leading to their development of food allergy. In the case of her son, Elizabeth also blames herself for agreeing to the antibiotic treatment in the first place. Because of what happened to Viola, she says she should have known better.

"I sorely regret it to this day," Elizabeth says, "because I am convinced that's what caused Brian's leaky gut and a litany of other issues."

Part of Elizabeth's rationale for thinking that early exposure to antibiotics caused Viola's and Brian's conditions is that no one else in her family has ever developed any allergies whatsoever. In fact, allergy is so rare in her family that her parents initially didn't believe the diagnosis. They argued that "back in their day" everyone ate everything and was fine; according to them, food allergy was a bunch of stuff and nonsense. But when both Viola and Brian ended up in the ER on multiple occasions for anaphylactic reactions to foods, it became all too clear to Elizabeth's parents that their grandchildren's allergies were indeed "real."

Since her children developed multiple allergies, Elizabeth's family's routines have been severely affected. "My life revolves around cooking for them," she explains. "We don't eat out. We don't trust people preparing their food." Instead, Elizabeth gets up each morning at six-thirty to cook a special breakfast, one that avoids everything that all three children are allergic to. Then she cooks and packs their lunches: "Every morning I have twenty-four holes to fill, twenty-four sections of lunch boxes and snack boxes to fill to get them to school." She cooks everything from scratch, because most packaged foods contain at least one ingredient that one of her children will react to.

A few months ago, they went on vacation with four other families and rented an Airbnb. Brian ended up in the ER with anaphylaxis due to cross-contamination. Elizabeth says they will never share a house in which she isn't the "cleaning boss" ever again.

Even though he's only a toddler, Brian, whose allergies are the most severe, knows that some foods are dangerous.

"I'll ask him, 'Do you know why you can't have that?'" Elizabeth says. "And he'll say, 'Yes, Brian allergic. Makes me owie. Mommy give me shot and we go to hospital.' He remembers the EpiPen. He remembers it because those things hurt. It's an inch-and-a-half needle jabbed into you."

Brian runs away from Elizabeth every time he sees her pack an EpiPen into one of their bags. She says it makes her feel like she's the biggest monster in the world, and she can't help but feel guilty. Not just because Brian associates bad feelings with the EpiPens she has to carry, but because she ultimately feels responsible for his allergy.

Elizabeth's story—and guilt—is not unusual. Many caretakers of children with allergies wonder, and worry, about why their children developed the condition in the first place. They fear, not without some evidence or solid reasoning behind it, that something about their lifestyles, household environments, or habits has contributed to their children's suffering. The parents of children with severe eczema and food allergy, especially, often perform a kind of archaeology of their past, sifting through their memories for any repeated actions or early exposures—anything that might help them make a bit more sense out of an ostensibly senseless situation. I understand this impulse.

Much like Elizabeth and other moms in her support group, a lot of regular people I talk to or interview about allergies have their own theories about what's causing our allergies, most aligning in one way or another with the top scientific candidates, and most having something to do with the idea that changes to our environment are likely to blame. In a demographically representative survey of eight hundred Americans I conducted in September 2018, nearly 57 percent of respondents thought that pollution was to blame for allergies. Another 48 percent thought man-made chemicals were likely involved. Tying for third place, at 38 percent each, were climate change and changes to our living and eating habits.

When I began doing serious research for this book, it seemed intuitively obvious to me that a multitude of environmental changes—like pollution or decreased exposure to bacteria, viruses, and parasites—coupled with the stresses of modern, urban life, were likely the dual keys to solving the puzzle of our rising allergies. Something about our relationship to our surroundings, I reasoned, was out of whack. I felt certain about this because I thought I had concrete "evidence" to support my theory: I had been a happy, healthy, nonallergic person before I had started living in dense, pollution-riddled cities like New York, San Francisco, and Hong Kong.

My early childhood in rural Indiana had been, at least in my recollection of it, nearly idyllic. We ate vegetables and fruits grown in our own pesticide-free gardens. We spent our days outside in the fresh country air. We played in the dirt among rows of corn and in neighbors' barns. We ate clover leaves and dandelion stems and, sometimes, grass from our yards. In short, I had bought into one of the main theories for why allergies have been rising for the past two hundred years: the hygiene hypothesis.

The hygiene hypothesis posits that people who are "too clean" develop allergies, that encountering a variety of germs very early in life (before you're a year old), particularly on farms or as a member of a large family with multiple siblings, is protective. Proponents of the hygiene hypothesis believe that exposure to a little "dirt" is good for us. Encountering the right germs at the right time trains the infant immune system to respond correctly to a range of external stimuli. If it doesn't get this initial training—or gets the wrong exposures, or the right exposures at the wrong times—then it is primed to overreact later in life.

By the logic of the hygiene hypothesis, I should have been correct in my assumption that my large farm-country family (most of my aunts and uncles had three or more kids) would have a much lower rate of allergies. Yet when I called home to ask about allergies in our extended family, I was quickly reminded that my family's immune systems are just as irritated as any urban family's. Rural life, then, at least going by my family's history of allergies and some of the latest research (which we'll delve into in a minute), might not be a panacea for our immune system development after all. In other words, the hygiene hypothesis might not be the final answer to the riddle of allergies.

But as we've already seen, it isn't just people living in the twenty-first century who surmised that alterations to their lifestyles and surroundings were leading to trouble. In the era of John Bostock and

Charles Harrison Blackley, hay fever was blamed on changes to agricultural methods along with rampant air pollution. As early as 1951, Dr. Walter Alvarez, a prominent physician who wrote a syndicated medical column for newspapers across America, was already blaming increased chemicals in the environment for the dramatic rise in respiratory allergies and asthma.

For the past two hundred years, we've been collectively fretting that our allergies are a symptom of a much larger problem: that something we are doing or have done is making all of us very, very irritated, itchy, uncomfortable, and sick. It's the idea that ties all the hypotheses we'll explore in this chapter together. I call this the "we're doing this to ourselves" theory of allergy causation. It's almost intuitive for us to think that changes we've made to our collective way of life are causing allergies to get worse. But are we right? When I interview allergy experts, I force them to come down on a side. Many say the hygiene hypothesis is correct—and it remains one of the front-runner theories. But as we'll explore throughout this chapter, many others think it's our diet, that changes in the way we grow and prepare food have altered our gut microbiome, fueling allergies. Still others argue that the various man-made chemicals and plastics we come into contact with on a daily basis are making our immune systems more irritated. What everyone can agree on is that our gene-environment interactions (also called epigenetics) have a large role to play in the rise of allergies, as does the makeup of our nose, gut, and skin microbiomes.

We're about to explore the major theories of allergy causation that focus on the effects of our "modern" lifestyles on our immune system function. The ways in which we produce, prepare, and eat food; the modern work culture with its continuous lack of sleep and high levels of stress; the antimicrobial agents, antiparasitics, and antibiotics that we use in human medicine and feed animals; gardening and our obsession with having a lush, grassy yard—all of these are suspects in the development and steady rise of allergies. While blame for allergy may have shifted from neurotic behaviors and anxious personalities in the nineteenth century to our diet and microbiome in the twenty-first, our culture and daily habits have come under steady scrutiny for the roles they may play in our increasing irritation for more than two hundred years. When all is said and done, we *are* right to blame ourselves, at least partially. Our entire modern lifestyle is likely at the root of the recent rise in allergies.

WHITE, WORRIED, AND WELL: A BRIEF HISTORY OF BLAMING ANXIETY AND STRESS FOR ALLERGY

In the 1800s, before it was clear exactly what was causing hay fever or asthma, doctors often blamed the patients—at least to some degree—for their worst symptoms. In 1859, Dr. Henry Hyde Salter, an early British researcher and himself an asthma sufferer, felt that hay fever and asthma were primarily nervous diseases. Dr. George W. Bray, an allergist working in the allergy clinic at the Hospital for Sick Children in London in the early part of the twentieth century, suggested that "many allergic conditions follow immediately on fears and emotions, whilst anticipation has a pernicious effect."[1] Research conducted at Harvard at the turn of the twentieth century suggested that asthma might be caused in children by harsh punishment, or by a "fixation or hate subconsciously directed toward the mother."[2] These views of allergy causation were not that uncommon and were linked to the type of patients most allergy researchers were used to seeing in their clinics.

The first respiratory allergy sufferers, or at least those going to see physicians for their ailments, tended to be white, urban, and educated. Many of them were young boys and women, and thus doctors began to associate physical frailty or weakness with hay fever and asthma. Early scientific texts, even those published as late as 1935, defined allergy as a "hypersensitivity" or "hyper-irritability or instability of a portion of the nervous system." In other words, nervous, anxious people, or "neurotics," developed allergies. Trigger mechanisms for hay fever and asthma were thought to be not only the allergen itself but also anything that upset the nervous system of the patient and brought them out of an otherwise "balanced allergic state."

Writing in 1931, Dr. Warren T. Vaughan, a famed allergist, argued that any stressor could impact this "equilibrium" and cause an allergy or asthma attack: an infection, insomnia, anxiety, hormones during menstruation or pregnancy, emotional upset, or physical activity.[3] His contemporary, Dr. Samuel Feinberg, writing in 1934, suggested that allergy patients tended to have higher than average intelligence but were also more emotional and "temperamental" with a more "alert nervous system."[4] In 1939, the chief of the allergy clinic at Lenox Hill Hospital, Dr. Laurence Farmer, argued that the psyche most certainly played a "decisive role in the allergic drama" and that emotions often triggered severe attacks.[5]

As an allergist of some repute and decades of experience, Dr. Arthur Coca suggested in his 1931 book that "overfeeding" and "underexercise" could cause asthma attacks.[6] He identified common changes to his patients' personalities prior to the development of their symptoms. Dr. Coca argued that "irritability and crankiness are common forerunners" for food allergies, and that a type of observed, general "nervousness" might be the sole symptom of their malady. However, he also noted that "psychic treatment," or psychoanalysis, was not particularly useful in managing allergies or preventing attacks.

Dr. Albert Rowe, a food allergy specialist who wrote a canonical book on the topic in 1931, argued that many people didn't take food allergy seriously, blaming the patients for their malaise. Because their self-reported symptoms were largely invisible to the naked eye, food allergy patients faced suspicion and increased scrutiny. Rowe thought that food allergy was likely widespread, if only physicians would take more notice of it as a legitimate medical condition. He complained that "many physicians think that idiosyncrasies to foods are imaginary,"[7] and noted that females were more likely to present with food allergies, which may have been why more doctors were dismissive of the symptoms. Women's symptoms were viewed by many doctors at the time as mere "fancy."[8] Rowe urged his peers to be more open-minded and to cultivate a willingness to try food elimination diets on patients who were incurable via other means and methods.

In the 1950s, Alvarez, writing for the Mayo Foundation, suggested that nervous or emotional strain could trigger an allergic reaction or sensitize someone to a particular allergen, causing either respiratory or food allergy. He noted that it was unclear whether or not stress alone could produce "allergy-like" symptoms, and that it was a topic of debate among physicians and specialists. That being said, it was clear that allergy was a problem in highly sensitive, well-educated people and that the act of overeating could produce an allergy to certain foods in these individuals (at the time, this idea still seemed plausible, though now we know it to be patently false). Going further, Alvarez suggested that it was difficult, if not impossible, to determine whether or not a patient had a food allergy, was depressed, or simply had a "prejudice against some food."[9] By 1953, Feinberg was counseling in a pamphlet on allergy that many instances of so-called food allergies were nothing more than cases of depression, insomnia, or fatigue.[10]

Some of the earliest treatments for allergies reflected this easy association between a patient's mental state and their worst symptoms. From the nineteenth century until the early twentieth, many physicians counseled their hay fever and asthma patients to avoid any stressors or physical exertion that might trigger an attack, and regularly used opium, alcohol, or other sedatives to treat the conditions. Though this practice gradually died out, in part due to its ineffectiveness and the increasing recognition that opiates could be very dangerous to prescribe, most medical books listed sedatives as a viable treatment option for tough allergy cases until the 1960s, when allergy advice still retained echoes of the association between stress and emotions and bouts of allergy. A pamphlet published by the Allergy Foundation of America discussed the relationship between emotions and attacks of allergy, writing that "excitement, anger, even fear, can trigger allergic attacks."[11]

More than a century of equating neuroticism and stress with allergy led many allergy sufferers and non-sufferers alike to believe that allergy and asthma were a weakness of urban, wealthy elites, or the "white, worried, and well." (And, unsurprisingly, we'll see this conflation again when we examine the cultural understandings and media representations of allergy in the twenty-first century in chapter 10.) By 1947, the celebrated immunologist and allergy specialist Robert Cooke decried the devolving of allergy into a replacement diagnosis for neuroses, developing a bad reputation among "serious" medical practitioners as nothing more than a fad or fashionable diagnosis.[12]

This stigma and personal blame have lingered, popping up in our modern mistrust that people with allergies might be "faking it," or the belief that allergies are not a "serious" illness like cancer or diabetes. (We'll look more deeply at this issue in chapter 10.) It echoes, too, in the ways in which allergy sufferers—especially eczema patients— often connect daily stressors and their overall mental health to an increase in allergic reactions or worsening symptoms. Many people I talk to about their allergies feel that the relationship between their allergy and their mental and physical well-being is nearly synonymous. The correlation goes in both directions; if they are healthy and happy, their allergies are less severe and attacks are less frequent. Stress and fatigue can be both products and causes of an allergy attack.

And, let's face it, we live in stressful times. This decade began with a global pandemic and some of the biggest wildfires, droughts, and floods we've ever seen. The global economy is still reeling from the effects of COVID-19 and is showing signs of a slowdown. Are the increased levels of anxiety and pressure we're all experiencing affecting our allergies? Is there a direct connection between our level of stress and our immune system? In short, the answer is yes, of course.

In the past few years, researchers have found evidence that stress directly impacts our immune responses via the histamine released from mast cells throughout our bodies. When we are under mental or physical pressure, our bodies release stress hormones, such as cortisol and adrenaline. Recent research performed at Michigan State University found that mast cells are highly responsive to one such hormone, called corticotropin-releasing factor (CRF1).[13] Researchers found that mice with normal CRF1 receptors showed an increased number of mast cells and mast cell degranulation, or histamine release, after they were exposed to CRF1. Mice that lacked the CRF1 receptor have far less activation of mast cells and as a result fared much better (the mice exposed to allergic stressors had a 54 percent reduction in allergic disease). In other words, the mice that were more prone to stress were also more prone to allergic responses—and their stress hormones directly activated their histamine responses. In a study of seventeen hundred Germans, researchers at the Technical University of Munich found a correlation between allergy and common mental health disorders.[14] Study participants who had perennial allergies were more likely to also suffer from depression; if they had seasonal pollen allergies, they were far more likely to have anxiety.

That makes sense to Dr. Pamela Guerrerio, a food allergy expert at the National Institutes of Health. When I interviewed her about the causes of food allergy, she mentioned the link between our use of proton pump inhibitors, a class of commonly prescribed medications that control the level of stomach acid (antacids), and IgE sensitivities to foods in adults. The medications lower the acidity of our gut, allowing food in our stomachs to get absorbed in a more immunologically intact form. Since the level of stomach acid is connected to our diet and our stress level, this is a clear-cut example of how our modern lifestyles are impacting our immune systems.

But it's not just food allergy patients who might be experiencing the negative effects of stress. Dr. Peter Lio, an eczema expert, told me that the link between stress and the skin is obvious. And the treatment for conditions caused—at least in part—by stress cannot be just more medications. It has to be, he argues, more holistic and encompass every aspect of the patient's lifestyle.

"You can actually show that if you stress somebody out, their skin barrier starts breaking down," Lio explained. "Even in a healthy person. Sometimes Western people are like, 'Oh, come on, quit talking about stress.' But this is physiologic stress damaging the skin. This is a real thing. And we live in an incredibly stressful society."

It seems perfectly clear, at least according to more recent scientific findings, that when our individual stress levels climb, our allergic responses go into overdrive. This is different, however, from the early days of allergy medicine, when a patient's mental state was often directly blamed for their allergies. What twenty-first-century allergy experts like Guerrerio and Lio are arguing is that their patients' lived environments—their workplaces, homes, cities, towns, and communities—are sites of outside stressors.[15] Long hours, less affordable childcare, smaller social circles, a bad economy, lengthy commutes, more overtime: Any and all of these things can cause a patient's stress levels to climb. And increased stress is making us all a lot more irritated.

THE HYGIENE HYPOTHESIS EXPLAINED

As the last century rolled on, and more research on immune system function piled up, the focus on the etiology, or cause, of hypersensitive immune responses shifted from inheritance, exposure to allergens themselves, and the patients' baseline personalities, to the microbial content of the modern lived environment. Let's look more closely at the hygiene hypothesis, perhaps the most well-known and most often espoused theory of allergy causation. You're likely already familiar with the idea that being "too clean," or too hygienic, is not all that terrific in terms of childhood development. Maybe you've heard that it's okay to let children play in the dirt, to get a little messy, to slobber on one another—that it's good for them. That's the basic idea behind the hygiene hypothesis, first posited to try to explain the explosion of asthma, eczema, and food allergies in the last half of the twentieth century.

In 1989, David Strachan, an epidemiologist, published a short article in the *British Medical Journal* (*BMJ*), titled "Hay Fever, Hygiene, and Household Size."[16] Strachan used data from a national sample of more than seventeen thousand British children born during the same week in March 1958. He looked at three things: (1) how many of the study participants self-reported symptoms of hay fever at age twenty-three; (2) how many of the participants' parents had reported hay fever in the subjects at age eleven; and (3) the parents' ability to remember if their child had eczema in their first seven years of life. Strachan examined many variables to explain the data, but the association that stood out to him, and the findings that he reported to *BMJ*, centered on the children's household size and their birth order.

What Strachan found, looking at that initial data, was that younger children seemed to be the most protected from developing hay fever or eczema, despite any differences in socioeconomic class. Strachan posited that the lowered rates of allergy might be explained "if allergic diseases were prevented by infection in early childhood, transmitted by unhygienic contact with older siblings, or acquired prenatally from a mother infected by contact with her older children." Smaller family sizes, improvements in housing, and higher standards of cleanliness might have combined to reduce the opportunity for children to become exposed to a wide variety of microbes. In other words, Strachan's findings suggested that mild childhood infections might be beneficial to the developing immune system.

At first this idea was rejected, since many immunologists still believed that bad infections could trigger allergy, especially asthma. But Strachan's ideas were eventually taken up and popularized after researchers discovered that IgE-mediated (or antibody-driven allergic) immune responses were driving many allergic conditions. It seemed plausible that a lack of early exposure to certain germs was the underlying problem, leaving the immune system "untrained" and hyper-responsive in later life. Early work on the microbiome and commensal bacteria (the friendly bacteria that live inside the human gut, in our nasal cavity, and on our skin) "led to a reformulation of the hygiene hypothesis as the 'old friends' or the 'biodiversity' hypotheses of allergy, which propose that changes in the environment, diet, and lifestyle associated with Westernized, industrialized countries have altered the diversity of the gut and skin microbiomes."[17]

The "old friends" hypothesis posits that humans are more at risk of chronic inflammatory diseases like allergy and autoimmune disorders because we no longer regularly encounter some of the microorganisms that we evolved alongside for millennia.[18] These "old friends," the theory goes, helped regulate our immune function. Their risk to human health was minimal, and a healthy immune system could easily keep them in check. Doing so trained the developing human immune system, making it more robust and adaptive to its normal environment.

The problem, from this perspective at least, is that in the absence of these old friends, our immune system doesn't get the early training it needs to better self-regulate. Instead, it overreacts to otherwise harmless stimuli like pollen or dust mites.

When taken together, these two closely related theories explain "the farmhouse effect." When I sat down with the renowned immunologist Dr. Cathryn Nagler at the University of Chicago, who does cutting-edge research on the microbiome and its relationship to allergy, she explained how the hygiene hypothesis and the idea of microbes as old friends combine to produce an almost idyllic conception of farm life. Farmhouses, with their tilled soil, muddy barns and stables, and fertile fields, come with a lot of bacteria, viruses, and parasites.

"There's a nice literature, an older literature, on how farm life is protective," Nagler said. "Diversity is good for the microbiota. All of the microbes that colonize us later in life come from the environment."

As Nagler explained, if you alter the environment, you alter the microbiota. If you have better sanitation, move away from farms, and have fewer children, then you cut off your supply of a richly diverse microbiota. You become, in essence, less intimate with microbes in your day-to-day life. And intimacy with friendly germs, especially in the first few years of life, does seem to be protective of a wide variety of immune disorders—but not all of them. The hygiene hypothesis relies on the argument that "too clean" environments skew our immune function toward T helper cell type 2 (Th2) responses—or allergic responses—mediated, at least in most cases, by IgE. (As you might recall from the first chapter, IgE are antibodies formed by B cells to fight off specific things that T helper cells have encountered before.) However, Th2 disorders like seasonal allergies are not the only immune conditions on the rise. In recent decades, we've seen T helper cell type 1 (Th1) diseases, or autoimmune disorders like multiple sclerosis, increasing as well. There is a lot of scientific evidence to back up the hygiene hypothesis and the so-called farmhouse effect, and most of the allergy experts I spoke to find the theory very compelling. But, as we've seen so many times before, allergy causation is complicated, and the hygiene hypothesis can't explain everything.

Recent studies have suggested that there is a measurable farmhouse effect, but researchers are uncertain about *which exposures* are protective and what mechanisms they might be triggering to produce that protective effect. What seems certain is that exposure to livestock from early childhood dramatically lowers the risk of developing all allergic conditions later in life. In particular, exposure to stable dust seems to prevent most allergic responses.[19] Something in "farm dust" is effective—bacteria, viruses, fungi, or even more allergens themselves—but it's not entirely clear which components of the dust are protective, and which aren't. Another study of rural areas in Austria, Germany, and Switzerland showed that a farming environment was more protective against hay fever, atopic sensitization, and asthma.[20] If infants spent a lot of time in stables and drank cow's milk in the first year of their lives, then their rates of allergic diseases were dramatically lower *even if* their IgE results showed some sensitization. In other words, they might have an underlying sensitivity to some allergens, but that sensitivity did not develop into full-blown allergic responses.

In a different study that looked at the immune function of mice raised in a lab versus in a barn on a farm, the farmhouse effect was strongly supported.[21] The results of studies in mice are, in fact, one of the key supports for this theory. Dr. Avery August, an immunologist at Cornell University, explained to me that pathogen-free mice bred for laboratory studies have dramatically different immune systems than do their "unclean" peers; they have immune systems that resemble a human newborn baby's immune system. When you place those "clean" mice in a "dirty" environment—like the mice study did to simulate farm life—their immune systems change to look more like that of an adult human.

This tracks with research in humans that suggests that germ-ridden environments other than farmhouses can also be protective against allergies. Children and adults who live with dogs have lowered rates of both asthma and obesity, likely due to more indirect exposure to bacteria that dogs carry and track into the home.[22] A recent NIH-sponsored study showed that exposing infants to high indoor levels of pet and pest allergens (cockroach, mouse, and cat allergens, to be specific) lowers their risk of developing asthma by the age of seven.[23] But exposures to bacteria might be both protective and not—all depending on *which* bacteria.

Consider the fascinating case of *Helicobacter pylori,* or *H. pylori,* a common gut microbe. You may know *H. pylori* as the culprit behind gastrointestinal ulcers, chronic gastritis, or even some forms of cancer.

Although the species *H. pylori* was discovered by scientists in 1982, there is speculation that our colonization by the bacteria is much older (taking place around sixty thousand years ago) and required repeated contact in small, close-knit groups—that is, the way humans typically lived until fairly recently. There are many different strains of *H. pylori,* and their prevalence in humans was estimated to hover around 80 percent until after World War II, when the introduction of prescription antibiotics like penicillin to treat common infections led *H. pylori* to begin to disappear from the human gut. Today, it is estimated that around 50 percent of all humans are infected with *H. pylori,* with rates hovering as high as 70 percent in some African nations and as low as 19 percent in some European nations.[24]

This is in line with the hygiene hypothesis, since transmission of microbes is far easier in large, crowded households with many siblings. *H. pylori* is usually acquired in early childhood, after the first year, and is transmitted via fecal-oral, oral-oral, or vomitus-oral routes. In the absence of antibiotics, *H. pylori,* once acquired, can persist in the gut for decades, often for the entire life of its human host. Most people living with *H. pylori* have no symptoms or ill effects.

The stomachs of people with and without *H. pylori* are immunologically different and there is speculation that people with *H. pylori* have a larger gut population of regulatory T cells (Tregs). That's important because Tregs play a crucial role in tamping down our inflammatory immune responses. Although infection with *H. pylori* is associated with having more immune cells in the gut, some researchers have proposed that may be a normal, rather than a pathological, response to the bacteria.[25] In other words, *H. pylori* may be beneficial in some situations. In fact, people who lack the bacteria are much more likely to suffer from gastroesophageal reflux disease (GERD), or acid reflux, and there is evidence that *H. pylori* plays a protective role against childhood-onset asthma. Because of this, some researchers conclude that *H. pylori* is likely an "amphibiont," or a microbe that can be "a pathogen or a symbiont, depending on the context."[26]

All of this suggests that there is credence to the basic premise of the hygiene hypothesis: that we need regular exposure to friendly bacteria to train our immune systems. However, it's likely not that simple of an equation that living with more and more diverse microbial populations automatically produces a better immune system function. Dr. Thomas Platts-Mills, the director of the Division of Allergy and Clinical Immunology at the University of Virginia School of Medicine, has argued that the hygiene hypothesis cannot possibly explain the rise of allergies. It can't be, he told me, the culprit we're looking for—at least not by itself. His argument relies on our more recent history of "cleanliness."

Throughout the twentieth century, hygiene standards were more widely adopted. Improved sewage systems and potable drinking water meant that human exposure to microbes was far less frequent, at least by way of ingesting them. Regular infection by helminths, or intestinal parasites, had decreased due to food and water quality controls and the increased wearing of shoes. Since this was a period of people moving from rural farms into urban centers, the general population also saw lower exposure levels to farm animals and decreased diversity of the bacterial populations they were regularly exposed to on those farms and in the soil. Family size had also decreased, perhaps contributing to less exposure to germs in children, but, as Platts-Mills is quick to point out, all of these changes were already complete by the 1920s, which doesn't explain the dramatic rise of asthma and allergic rhinitis beginning in the 1940s into the 1950s. Dr. Platts-Mills argues that the best explanation for the rise of hay fever and asthma is not the hygiene hypothesis alone but more likely "an increase in sensitization to indoor allergens and the loss of a lung-specific protective effect of regular deep inspiration." In other words, outdoor recreation was likely more protective against allergies than spending hours playing *Minecraft* or *Fortnite*.

If the hygiene hypothesis or farmhouse effect was correct, one would also expect to see a marked decrease in allergy rates in rural farming communities. Yet Dr. Jill Poole, chief of the Division of Allergy and Immunology at the University of Nebraska Medical Center, has found that around 30 percent of Midwestern farmers suffer from allergic disease directly linked to their agricultural lifestyle. Dust from grain elevators and animal barns, pesticide exposures, and grain rot from flooding cause what is colloquially called Farmer's Lung. So, while some farm exposures seem beneficial, others are clearly not.

And if family size, rural life, and socioeconomic status are linked in the original hypothesis theory, then one might expect that countries with larger family size, more rural populations, and lower socioeconomic status would have a lesser burden of allergic diseases. Yet in places around the world with larger families, a larger percentage of rural populations, and more families living at or below the poverty line, allergies are steadily increasing. One recent study found that half of Ugandans living in the capital city of Kampala have some form of allergy.[27] The study also showed that allergies are on the rise in rural areas in Uganda, although more urban dwellers are able to report to hospitals with symptoms of asthma, nasal congestion, or skin rash. Many Ugandans are self-treating with over-the-counter antihistamines, steroids, and antibiotics. Dr. Bruce Kirenga, a Ugandan allergy expert, said he thinks environmental pressures such as air pollution, rather than urban lifestyles, are to blame.

All of these findings taken together suggest that the farmhouse effect or the hygiene hypothesis might not be the smoking gun we're searching for. Although the theory makes intuitive sense, we simply don't have enough scientific evidence to definitively say that rural life, with its "dirty" or microbially rich environments, can fully protect us from allergic disease. And yet, the basic idea that *something* about our interactions with the microbial world around us has shifted as a result of our lifestyles and daily habits remains compelling. The hygiene hypothesis, then, is likely *partially* correct. Evidence is stacking up that some of our habits (particularly in relationship to our diets and food production) might be behind the recent rise of allergies—especially food allergy.

THE MICROBIOME AND FOOD ALLERGY

If you want to better understand how our modern lifestyles might be behind some of our biggest problems with allergy, particularly when it comes to our food production methods, our diets, our use of antibiotics, and our childbirth practices, you end up sitting across a table from a diminutive, deeply intelligent, empathetic, blond-haired woman named Cathryn Nagler—Dr. Nagler to her students, Cathy to her friends and colleagues. Everywhere I went to interview allergy experts, I heard her name uttered in the same sentence as food allergy. I quickly learned that this is because Nagler is one of the best immunologists in the world. Her research primarily focuses on the role our gut microbiome plays in the development of food allergy in children. She's been doing this work for a couple of decades now, and she remembers when food allergy rates first began to climb in the late 1980s.

"I saw it myself," Nagler says, pulling up some graphs on the computer screen tilted toward me. We're sitting in her office at the Univer-

sity of Chicago on a bright, sunny spring afternoon. "I have kids that
are twenty-three and twenty-seven, so I followed this in real time because cupcakes were excluded from the classrooms as my kids went
through school. Right around the late eighties and early nineties,
when food allergy rates were starting to increase, the American Academy of Pediatrics said to withhold peanuts and allergenic foods from
pregnant mothers, from nursing mothers, and from children with risk
of allergy until they're four years old. That was exactly the wrong
advice, and that fueled the fire and caused even more increase. Now
all of the push is for early introduction."

What Nagler is indirectly referencing is the now-famous Learning
Early About Peanut Allergy (LEAP) study, conducted by researchers
in the United Kingdom and the United States, led by Dr. Gideon Lack
at King's College London and published in *The New England Journal
of Medicine* in 2015.[28] The study found that decades of erroneous
advice to parents to avoid giving children younger than three years
old anything containing peanuts had led to a massive increase in the
incidence and severity of peanut allergy. Infants enrolled in the study
(four to eleven months old) were randomly assigned to two groups:
Parents in one group would continue to follow the advice to avoid
peanuts; parents in the other group would be told to introduce peanuts to their children right away. Infants in both groups were given
skin-prick tests for sensitivities to peanuts. Among those who tested
negative, the prevalence of peanut allergy at sixty months of age was
13.7 percent in the peanut avoidance group and merely 1.9 percent
in the peanut consumption group. Among those who had tested positive for sensitivity to peanuts, the prevalence of peanut allergy was
35.3 percent in the avoidance group and 10.6 percent in the consumption group. A recent study conducted in Melbourne, Australia,
found that changes to dietary advice on peanuts in 2016, following
the success of the initial LEAP study, had led to a 16 percent decrease
in peanut allergy among infants.[29] It's perfectly clear that introducing
peanuts to infants has a protective effect.

Nagler understands why parents might be hesitant, however, to
introduce allergens into the diet early. After all, why would anyone

trust the same people who gave them the wrong advice just a few years ago? Plus, she doesn't think there's definitive evidence that early introduction is good.

"You can be sensitized even before the first introduction of solid food," Nagler explains. "Kids get allergic responses within the first month of life. That means they could have been sensitized by breast milk or by the skin. If you give early introduction to a kid like that, that kid is going to have an allergic response. So early introduction is risky, but now we know withholding is not good either."

So where does that leave us? Nagler is more worried about how our immune systems get sensitized in the first place. How does the body learn to tolerate some foods and begin to react negatively to others? She is convinced that food allergy as a phenomenon is part of a generational change.

"People will tell you that there is no history in their family of this," she explains. "Your allergy can develop at any point in your life. It used to appear between the ages of two and five. Now we're getting a lot more adult-onset food allergy. It used to be that milk, eggs, wheat allergy were outgrown. Now they're lasting into adulthood."

In other words, things have shifted. A lot. And not for the better. Food allergies are a signal of a larger problem.

Nagler shows me slides of the various changes, and I am scribbling down facts as fast as I can. She talks quickly, in part because she has so much to tell me. She runs through the various theories of allergy causation, like the hygiene hypothesis, and then she stops on a slide showing all the things that are likely contributing to our immune system's malaise: diet, C-sections, changes in food production, breast-feeding.

"The idea is that modern industrialized lifestyle factors have triggered shifts in the commensal bacteria," Nagler says. "Commensal bacteria" is a fancy way of referring to all the so-called friendly bacteria we live among and that exist within us and alongside us. "Inflammatory bowel disease, allergies, obesity, autism—all noncommunicable chronic diseases. They've all been linked to the microbiome."

And there it is: Nagler's answer to the all-important question of why allergies are on the rise. Changes to the makeup of our microbiome—or all the bacteria and viruses that live in our gut, helping to process our food into usable fuel for our cells—are driving changes to our immune function.

Recent studies have highlighted the connection between our diet, our use of antibiotics, and our gut bacteria in the development of allergies. A 2019 study showed that the gut of healthy infants harbored a specific class of allergy-protective bacteria not found in infants with cow's milk allergy.[30] This was followed by a study at Brigham and Women's Hospital in Boston that found that five or six specific strains of gut bacteria in infants seem to be protective of developing food allergies. A lead researcher on that study, Dr. Lynn Bry, surmised that our lifestyles are, for either better or worse, capable of "resetting the immune system."[31] Another study found that consuming higher levels of cheese in our diet may accidentally worsen allergy symptoms because the bacteria in some cheeses produce histamine—the naturally occurring compound that helps to trigger an effective immune response.[32] Researchers at the University of California, San Francisco, discovered a link between three species of gut bacteria and the production of a fatty molecule called 12,13-diHOME.[33] That particular molecule lowers the number of Treg cells in our gut, cells that, as we've already seen, are crucial in keeping inflammation at bay. The researchers found that babies who had higher levels of these three bacteria were at an elevated risk for allergy and asthma.

Nagler explains, "There are many, many immune cells in the gut. The gut seems to be microbiome headquarters. There's the most diversity there, certainly the greatest numbers, especially in the colon. That goes up to the trillions."

Ultimately, most of us living in the twenty-first century have changed our microbiome makeup. Our diets are the real culprit, according to Nagler. When we go from eating foods with a lot of fiber to consuming highly processed foods that are loaded with sugar and fat, we end up starving beneficial bacteria in our guts. We don't give them the food that they need.

"We've coevolved with our microbes," Nagler says. "Now we are missing their food. They can't live without their food."

There's also the use of antibiotics that kill off not only the bacteria that cause our strep throat and sinus infections but also our gut bacteria. And we eat meat from animals that have been given low-dose antibiotics to make them fat. Nagler surmises that all of this is likely having a big effect on our own microbiome. We're experimenting on ourselves, she said, to deleterious effect.

Nagler has developed a new theory called the "barrier regulation" hypothesis. In essence, our gut and skin microbiomes regulate what is allowed into the body and what is kept out. Commensal bacteria on the skin and in the gut are integral to maintaining barrier function. Nagler explains that a single layer of epithelial cells is all that stands between us and everything around us, making sure that everything that enters our bodies is either inhaled or ingested.

Indeed, researchers have recently discovered a link between a gene coding for an antiviral protein in the gut, changes in the gut microbiota, and greater intestinal permeability and severe allergic skin reactions in mice.[34] Gut microbiomes are an intricately balanced mix of different species of bacteria, viruses, and fungi. Mice lacking the gene for the antiviral protein had a changed microbiome (in layman's terms, the amounts and types of different bacteria and viruses shifted significantly). This suggests that our immune systems have developed ways of coping with microbes in our gut and keeping things in balance. When the composition of the microbiota changes, the immune components' various responses shift, making us more miserable in the process. This is evidence of how genetics (the gene) and the environment (changes to the gut microbiota) interact to produce allergy, but also proves Nagler's larger point that altered gut microbiota can have a direct effect on allergy.

Remember August's description of human immune cells as curators of the human body? The barrier regulation hypothesis dovetails nicely with the conception of our immune system as a whole—microbiome included—as curator of what can and cannot be a part of us. Without the regulation that those barrier cells provide, entire proteins can pass through our skin or our gut into the bloodstream, where they encounter our immune cells. The allergic person's immune system is wholly functional; it is simply doing the job it was meant to do. The ultimate problem, at least from Nagler's point of view, is that it's being asked to perform a job different from the one it was initially trained to do. So, from this perspective, allergic disease is a barrier problem, not necessarily an immune system problem.

All creatures, even invertebrates, have an associated microbiota, Nagler explains, which perform vital physiological functions. Without microbiota, there would be no life at all. The human gut encounters antigens from a hundred trillion—or 100,000,000,000,000—commensal microbes and more than 30 kilograms, or 66 pounds, of food proteins per year. Cells making up the gut barrier have to discern between what are harmful—pathogens like damaging outside bacteria or viruses—and what are harmless antigens. As Nagler and her former student Dr. Onyinye Iweala, an immunologist at the University of North Carolina School of Medicine, argue in a recent review of the human microbiome's relationship to food allergy, "It is increasingly evident that a functional epithelial barrier engaged in intimate interplay with innate immune cells and the resident microbiota is critical to establishing and maintaining oral tolerance."[35] To put it simply, that means that a healthy immune response to food relies on an intricate balance between our epithelial cells, the friendly bacteria that live inside us, and the types of food that we ingest. Changes to any part of this equilibrium can spell big trouble, as we saw with Elizabeth's children at the start of this chapter.

From Nagler's perspective, Elizabeth's theory that antibiotics were ultimately to blame for her children's food allergies might not be that far-fetched. Changes in the gut microbiome in young infants and children can lead to a greater risk of developing allergic responses as the children get older. And it seems as though our children's earliest environments are the most crucial.

The microbiome has been shown to be incredibly stable by age three. Alterations before this age seem to be critical to the development of allergies, or not. A study at the Pasteur Institute in France found evidence in mouse models for the role of gut microbiota in the development of a healthy immune system as young as three to six months of age, the stage of development when most human babies are first introduced to solid foods. Bacteria in the gut increased tenfold to a hundredfold after solid foods were introduced.[36] This stage of rapid microbiome growth and development, which they call "pathogenic imprinting," seems to determine one's susceptibility to inflammatory disorders like allergy and autoimmune disorders in adulthood. Antibiotics could, in theory, disrupt this developmental stage, producing a greater risk of all allergic diseases.

So far, the scientific evidence appears to back this up.[37] Research conducted by Rutgers University and the Mayo Clinic found that children under the age of two who are given antibiotics are at greater risk for asthma, respiratory allergies, eczema, celiac disease, obesity, and ADHD. The study looked at 14,572 children born in Olmsted County, Minnesota, between 2003 and 2011.[38] If antibiotics were given in the first six months of life, the risk increased dramatically. Researchers found that 70 percent of the children in the study had been prescribed at least one antibiotic in the first forty-eight months of their lives (typically for respiratory or ear infections). Another recent study found that antibiotics can allow for the growth of non-pathogenic fungus in the human gut, which may enhance the severity of respiratory allergies.[39] Finally, a study of babies in Finland and New York found that C-sections and antibiotics both correlated with altered gut microbiomes and greater risk of allergies in infants.[40]

These findings don't surprise Nagler. In our interviews, she underlined that vaginal births provide the infant with what are known as "founder bacteria." As the baby moves through the vaginal canal, it is exposed to its mother's friendly bacteria. Then, breastfeeding introduces more helpful bacteria into the infant's gut.

"The bacteria colonize in an ordered ecological succession," Nagler explains. "The lactate-producing bacteria come in first. The next bacteria that come in are the ones that are expanded by breast milk. If you skip over both of those processes, which many people have done, you've disordered the microbiome. The first one hundred to one thousand days of life are absolutely critical for the development of the immune system."

Research has shown that babies born by C-section have not been exposed to the correct, harmless vaginal founder bacteria, but they have been exposed to potentially harmful hospital bacteria. One recent study found that lactobacillus-containing probiotics—the same bacilli found in breast milk—lowered SCORAD (SCORing Atopic Dermatitis) scores for children under the age of three who had moderate to severe atopic dermatitis, or eczema (although there was no measurable benefit for milder eczema). Breastfeeding for the first three months of life has also been linked with lower risk of respiratory allergies and asthma. In a study of 1,177 mother/child pairs, babies who were breastfed had a 23 percent lower risk of allergies by age six and a 34 percent lower risk of asthma (but only if there was no family history of asthma).[41] Breastfeeding, if not done exclusively, did not lower risk. If a mother supplemented her own milk with formula, the protective effect all but disappeared. (Important aside: If you're a mother and you're panicking a bit right now, please don't. There are many valid reasons to have C-sections and to choose formula over breast milk. We'll come back to this a bit later, but a lot of it is complicated, and there's much that we still don't know about these interactions.)

Nagler reminds me that the cattle industry has been giving cows low doses of antibiotics for years to make them fatter and more commercially viable. We also eat food that is low in fiber and highly processed,[42] with added sugars and fats.[43] That means that the foods we're introducing into our guts are different from the foods our ancestors were eating for millennia. And that, of course, will affect the types of bacteria that can flourish inside us.

Even something as simple as changing the sheets can change our microbiomes (we'll dive more into how chemicals are playing a role in the following section). Researchers from the University of Copenhagen's Department of Biology and the Danish Pediatric Asthma Center looked at samples from 577 infants' beds and compared them to respiratory samples taken from 542 of those infants at about six months old.[44] Researchers found 930 different types of bacteria and fungi. A correlation was found between bacteria in bed dust and those found in the associated children; while the two populations of bacteria were not exactly synonymous, they did seem to directly affect each other. An increase or decrease in respiratory bacteria mirrored an increase or decrease in the bacteria in the infants' beds. The research suggests that less frequent changing of bed linens may be beneficial to the health of our nasal and airway microbiome.

In essence, more diverse bacteria around us and inside us are overall a positive thing for our immune system function. In many interviews, I heard a longing on the part of researchers to return to a simple, less technologically driven, way of life. Much of that centered on which foods we consume, and how we produce them. One top allergist dreams of performing the ultimate control study to prove that our modern lifestyle and habits are negatively affecting our immune systems.

"Imagine," he said, "if we could get a group of people to revert back to a much older way of life. Eat foods grown without pesticides. Eat whole foods and a wide variety. Don't use dishwashers or detergents. Do you know what would happen? No more allergies. I just wish I could prove it."

A Brief Note on Our Diets and Nutrition

At this point, you may be hankering for more information about how we might be able to change our diets to help balance our gut microbiomes, and thus our immune systems. And while I understand this desire, I have to disappoint you yet again. There simply isn't enough scientifically valid evidence to support any dietary changes. I can, however, tell you a few things based on what we know now.

One: It doesn't help your immune system if you eat local honey. There is absolutely no evidence to support the theory that consuming honey that contains local pollen grains will help with your respiratory allergies. Local honey is yummy, though, so there is no harm in indulging your sweet tooth.

Two: Probiotics don't really work either. There's simply not enough evidence to support taking probiotic supplements for any allergic condition. They also don't help you to manage your gut microbiome. Many of the experts I interviewed wished you would simply stop spending your hard-earned money on them.

Three: Genetically modified organism (GMO) foods are not contributing to our malaise. Pamela Guerrerio told me that there is no data whatsoever that GMO foods are linked to the development of food allergy. Her rationale is sound. Food allergy has been around a long time, for centuries, long before the discovery of the double-helix structure of the gene in the mid-twentieth century. If GMOs were able to cause an allergy it would be because they were introducing new proteins to the immune system—but that would create a new allergy, Guerrerio argues. And we don't have any "new" food allergies—just more of the old ones.[45]

The good news is that scientists like Nagler are diligently trying to figure out which microbes are critical for healthy immune function—and they have some good candidates. But for now, we don't have any concrete techniques to alter our microbiomes in specific ways to aid our immune function. The best advice is still to eat a balanced diet with plenty of natural foods. Until the science advances, this is all we've got.

MAN-MADE CHEMICALS AND THE DOWNSIDE OF SCIENTIFIC ADVANCES

"Man's progress creates problems," wrote Dr. Samuel Feinberg, a leading allergist and the first president of the American Academy of Allergy, Asthma & Immunology, in a 1950s pamphlet on allergy.[46] Feinberg pointed the finger at human ingenuity as a significant cause of the increasing allergies in the developed world. All of our tinctures and dyes, our synthetic fabrics and new plastics, our lotions and eyeliner and lipsticks and shampoos were beginning to wreak havoc on the human immune system.

Several experts I spoke with mentioned man-made chemicals as one of the main driving forces behind our worsening allergies, especially in the effect they may be having on our skin's barrier.

Dr. Donald Leung, an immunologist and the head of the Division of Pediatric Allergy and Clinical Immunology at National Jewish Health in Denver, is one of the world's leading researchers of atopic dermatitis. In a conversation we had about the causes of skin allergy and eczema, Leung argued that we overuse soap, detergents, and products with alcohol on our skin. We routinely use harsh antimicrobial products on our hands and to clean our homes, instead of simple water and soap. Our efforts to sanitize our homes and ourselves increased throughout the COVID-19 pandemic, when antibacterial wipes were completely sold out for months. All of this can negatively affect our skin barrier, making it more likely we'll develop an allergic condition. At Northwestern University's Feinberg School of Medicine (named after Dr. Samuel Feinberg), Dr. Sergejs Berdnikovs, an immunological researcher, came up with what's known as the unified barrier hypothesis to explain the development of allergy. His idea is that the barriers throughout the body, from our genitals to our eyes, are regulated by a variety of hormones; if those hormone levels are altered in one location, then they weaken the epithelial barrier there, leading to a greater risk of allergic response. Dr. Amy Paller, also at the Feinberg School of Medicine, explained the barrier problem in relationship to atopic dermatitis.[47] In research she conducted on mice, when adhesive tape applied to their skin stripped them of their barrier and an allergen was applied, it produced atopic dermatitis. The barrier defect, according to Paller, "made them extremely exposed to the antigens." Relatedly, the dual-allergen exposure hypothesis for food allergies extends the barrier hypothesis to argue that exposure to food proteins through a weakened skin barrier along with early ingestion of higher doses of food proteins can lead to full-blown food allergy.[48] That means if you make a peanut butter sandwich and don't wash your hands and then you pick up your baby, you may be depositing trace amounts of peanut protein onto their skin. If your child's skin is "leaky," then that protein will seep into the skin. If the baby then eats peanuts, it can trigger a peanut allergy.

"All the things we're putting on our skin, or the things we're putting on our babies' butts, are probably not good for our barriers," Dr. Robert Schleimer said when we sat down in his office in Chicago to discuss allergy. Schleimer is the former chief of the Division of Allergy and Immunology at Northwestern's Feinberg School of Medicine and oversees cutting-edge research taking place on its campus. "There are all sorts of compounds, glycerol-based and others. Some are charged or acidic, and many are alcohols, and they are all probably disrupting the skin's barrier."

Then Schleimer told me a story about the 1960s. His first job was working for a company called the Tidee Didee Diaper Service for $1.70 per hour. His job was to collect all the used cotton diapers and bring them back to the laundry facility to be cleaned and repackaged for delivery. As he reflected on the barrier hypothesis, he pointed out that cotton is a natural fabric. Now we use plastic diapers with antimicrobial properties and apply creams to babies' skin to prevent rashes from those materials. And that's just *one* of the changes that might be exposing our children to more irritants.

"You have these very tough detergents made of rough chemicals that break things down," Dr. Kari Nadeau, the director of the Sean N. Parker Center for Allergy & Asthma Research at Stanford University, told me. "And initially that was seen as positive. Then they started to see that, wait a minute, all these people working in the plants that are making those detergents have breathing issues. The fact that you're putting proteases (enzymes that break apart proteins) into detergents, and the fact that these detergents are meant to clean clothes or clean skin, clean hair or clean dishes . . . they actually could harm our bodies."

During our discussion, Nadeau was adamant about the downsides of modern living, especially when it comes to all the chemicals we're exposing ourselves—and our children—to on a daily basis. She pointed to the recent rise in severe eczema. In the 1940s and '50s, the image of a "squeaky clean" home was heavily promoted by the very same companies that made these new detergents (like Dow Chemical).

"That turned out to be a problematic image," Nadeau said. "It turns out that the way my grandmother lived on the farm was probably the right way to do things: not using a lot of detergents, not bathing every day, making sure you were exposed to a little bit of dirt, being exposed to the outdoors."

In one recent study, researchers at Simon Fraser University found that young infants (zero to three months) who lived in a home where household cleaning products were used more frequently were far more likely to develop wheeze and asthma by the age of three.[49] Researchers noted that most of the infants spent between 80 and 90 percent of their time indoors—heavily increasing their exposure to these products. As Dr. Tim Takaro, one of the study's authors, noted, children take more frequent breaths than adults and, also unlike adults, they breathe mostly through their mouths. Breathing in through the mouth, rather than through the nose with its natural filtration system, allows anything that's contained in the air to penetrate more deeply into the lungs. Their hypothesis is that fumes from the cleaning products inflame the respiratory tract and thereby activate the babies' innate immune systems. The frequent use of certain household products—air fresheners, deodorizers, antimicrobial hand sanitizers, oven cleaners, and dusting sprays—seemed particularly harmful.

Exposures to the wrong chemicals before birth can be equally harmful to a developing immune system. A longitudinal study conducted on 706 pregnant women in France found that babies born with higher levels of cadmium in their umbilical cords were more likely to develop asthma (24 percent higher) and food allergies (44 percent higher).[50] Cadmium is a restricted heavy metal, but is used regularly in batteries, pigments, tobacco products, and coating for metals. The same study also found that higher levels of manganese, often used in the production of stainless steel, was linked to an elevated risk for eczema as the children developed. Another study found that higher concentrations of plasticizers (solvents added to materials to make them more flexible, or "plastic") equated to a greater risk of developing allergies.[51] Researchers measured levels of butyl benzyl phthalate (BBP), a common plasticizer used in the production of polyvinyl chloride (PVC, more commonly known as vinyl), in the urine of pregnant women and new mothers. They found that exposure to these phthalates during pregnancy and breastfeeding caused epigenetic changes to specific repressors for Th2 immune cells responsible for generating inflammation. Phthalates are classed as toxicants and can enter our bodies via the skin, food, or lungs, and they seem to be able to switch off genes via DNA methylation, a common biological tool that our bodies typically use during embryonic development. In other words, the man-made substances that we live among affect not only our own immune system functions but also the developing immune systems of our offspring in vitro.

While natural substances are no panacea for our allergies (and they can be deleterious, as in the case of cadmium and poison ivy, for example), certainly thinking twice about the products we're using in our homes and on our skin would be a good place to start. Our immune systems could clearly use a break.

VITAMIN D AND OUR SEDENTARY, INDOOR LIFESTYLES

Changes to our work and leisure habits might also be contributing to the recent rise in allergies. Whenever I talk to allergy experts, one of the theories they frequently mention is our modern tendency to remain indoors for wide swaths of time. Pediatric allergists, especially, are more likely to mention how much the daily lifestyles of children today have changed over the past fifty to one hundred years.

When I sat down with Pamela Guerrerio at the NIH, she quickly mentioned that we live our lives in a lot of shade. The lack of the sun's UV rays on our skin, which our cells need to produce vitamin D, means that we're producing less vitamin D. And lower levels of vitamin D are likely also playing a small role in overall allergy causation. Vitamin D has been found to be somewhat protective against allergies (though that evidence is debated), suggesting that our move indoors is harming us in unintentional ways.

Dr. Scott Sicherer, the director of the Elliot and Roslyn Jaffe Food Allergy Institute at Mount Sinai in New York City, was the first person I interviewed for this book and the first person who alerted me to the role that vitamin D might play in the development of allergies. He told me that both autoimmune disease and allergy disease tend to occur at higher rates the farther away a person lives from the equator. That single fact made immunologists think about the possibility that vitamin D was involved in immune disorders, since people are exposed to less sunlight at higher latitudes.

"But is that the whole story?" Sicherer asked me, fanning his hands out on the desk. "There might be fewer people engaged in farming lifestyles at those latitudes. There might be different exposures to different things in different regions of the globe. It's so complex that we just don't know."

His colleague Guerrerio agreed with this point. She remarked that people around the globe have starkly different diets, too, and that fact, combined with less sunlight, might be having a compounding effect on the immune system. She told me that it's likely that several factors are involved in allergy causation—including our indoor-prone lifestyles—and that several interventions will likely be necessary to reverse the effects on our immune system functions.

THE LITERAL CANARIES IN OUR ALLERGY COAL MINES

For my money, the most compelling evidence in support of the idea that our twenty-first-century lifestyles and man-made environmental changes have spurred the growth of our allergies is this: Our companion species of thousands of years—dogs, cats, birds, and horses—all get allergies regularly.[52] Other species—those that do not live either in our homes or alongside us—do not.

Our pets' symptoms are very similar to ours: sneezing, snoring, asthma, vomiting, and over-grooming in cats; skin eruptions and persistent scratching and grooming in dogs; coughing and wheezing in horses. And they likely have allergies for the same reasons we do. After all, their immune systems are exposed to the same panoply of natural and chemical substances. The top allergen in dogs? Dust mites. The top allergen in horses? Their human-packaged feed. Cats are often allergic to grass, tree, and weed pollens. Cats and dogs can also be allergic to human dander, since we shed skin, too. Sound familiar?

Since people all over the world are more willing than ever before to give their pets the best care possible, many owners spend a lot of time and money trying to eradicate allergy symptoms in their companion animals. The methods are the same as for humans: take antihistamines and steroids or undergo immunotherapy shots. The challenge is that we don't know just how big the problem is because we don't have good data on pet allergies or their incidence, as we do for humans. We know that they occur, but we don't know if rates are increasing, or if vets and pet owners are just getting better at recognizing the signs.

To better understand how and why allergies affect our pets, I drove to Ithaca, New York, to visit experts working on the problem at Cornell University's College of Veterinary Medicine. I sat down with Dr. Elia Tait Wojno in her office nestled among Ithaca's lush, rolling green hills. It almost feels like I'm on a farm, which is partially true, since the college keeps a lot of research animals and animal patients onsite. Dr. Tait Wojno's office is large, bright, and organized. We're sitting opposite each other at a desk, soaking up the last vestiges of the afternoon sun.

Dr. Tait Wojno started her career doing work on parasitic worms and immune responses. She explains that the immune response to parasitic worms is similar to the immune response during allergic reactions in both humans and dogs. (Of course, in the case of worms, those responses are protective and, in the case of allergies, the responses are the ones causing the miserable symptoms.) By studying the immune response to helminths (a type of parasitic worm) in dogs, we can learn a lot about the basic immune functions involved in allergy.

Working with dogs allows us to observe how allergies function in something other than a mouse model. For decades, mice have been the dominant research organism in the field of immunology. But, as we saw earlier in this book, mice are not humans and mouse models aren't always the best predictors of what will happen in a human body. That is why there's a growing interest among allergy researchers to move beyond mouse models for studying the disease. Since some larger animals like cats and dogs have natural allergic disease, they might be good models for learning about basic immunology across species as well as doing drug testing for allergic conditions.

"You can learn something about dogs by looking at humans, and you can learn something about humans by looking at dogs," Dr. Tait Wojno says. "You are looking at a naturally occurring disease happening in a very similar environment. My dogs sleep in my bed—they're exposed to many similar environmental stimuli."

Mice, on the other hand, are confined to the lab and live in very controlled environments. Mice are also typically genetically inbred. The dogs that Tait Wojno works with are all individual dogs, born the old-fashioned way. In fact, she works with breeders to enroll dogs into her studies. She underlines, emphatically, that the dogs in her study are treated like pets because they are. These are not lab animals; they live at home with their owners. This is an important detail that allows researchers to ponder which components of our shared, lived environments, habits, and medical practices might be affecting our companion species as well as us.

Allergies in our pets offer potential clues to solving the mystery of allergy. If we can understand early immune response in animals, then we might be able to better understand the basic early response in humans. And that's one of the things we really don't understand in any mammals—the immune system's initial reaction to something it encounters and the subsequent set of decisions it makes in response. Ultimately, my visit to Cornell convinced me of one thing: Our pets, like us, are the literal canaries in the figurative allergy coal mine. The fact that our intimate companions have allergies is a sign that something humans are doing is irritating all of our immune systems.

THE MYSTERIOUS RISE OF ALPHA-GAL ALLERGY

As we have seen so far, no single factor fully explains the increase in allergies over the past two centuries, but industrialization—and the environmental and cultural lifestyle changes that follow it—seems to play a key role. As of this writing, the highest rates of asthma are in English-speaking countries and Latin America; the lowest rates are in Eastern Europe, the Mediterranean, rural areas in Africa, and China. It takes immigrants from less wealthy countries between two and five years to develop allergies when they move to a wealthier country. Other immune diseases such as autoimmunity tend to show parallel increases. As the economy strengthens, our immune systems malfunction at greater rates.

Dr. Alkis Togias at the NIH underscored this connection: "Many things change as societies develop. There is no question that environmental exposures are related, and lifestyle is related, to what we see happening."

Dr. Scott Sicherer put it this way: "What you end up seeing in the more sophisticated view of allergy are these networks and lines tracking out from a lot of different nodes in the system that are interacting in different ways. You'd need a supercomputer to try to tease out what different influences are happening in the genetic and environmental pathways. So . . . it's complicated."

All of this is *complicated* because our biology itself is *extraordinarily complex*—and old. As we've seen, part of the trouble is that our environments and lifestyles are changing too much and too quickly for our slow evolutionary systems to keep up.

The other thing racing to keep up with all these complex changes is . . . the scientists themselves. The study of allergy causation is difficult and expensive.

"People have had something to say about where allergies come from ever since they started studying them," Dr. Steve Galli at Stanford reminded me. "And they've been wrong—for example, about withholding peanuts. As we learn more about allergy, the theory about what is causing it changes and reflects what's known at the time. So, I'd say without a smoking gun, you can't be certain who is pulling the trigger."

We don't have that single smoking gun. But we do have several smoking guns, all still firing simultaneously. We have, to mix my metaphors, a whole forest of smaller wildfires creating a vast smoke screen—one that stands between us and putting out the fire of allergies. To illustrate just how complicated causation can be, let's turn to the latest allergy on the block: so-called meat allergy.

I first heard about meat allergy during a dinner with colleagues. We were interviewing for a new position in our department and, as is traditional, we were hosting that day's candidate. As we looked over our menus, the wonderful woman we were wooing for the job, an anthropologist who works on water pollution issues in agriculture, mentioned that she couldn't eat any red meat.

"A few summers ago, I was bitten by a tick and developed an allergy to all red meats," she said, ordering the chicken. "It's not a huge deal for me, and I don't go into anaphylaxis. I just get really, really sick and break out in hives."

I was midway through my research for this book and was fascinated. I wanted to know everything. She and her partner had moved to Tennessee when she started a new job after graduate school. As an avid nature lover and a researcher on rural farms, she had spent oodles of time outdoors, roaming around the watersheds of rivers, walking the edges of farms where tall grasses met up with the cultivated fields. In other words, she was constantly in tick heaven. While it wasn't a shock for her to return home from her excursions to find a tick or two, her recent diagnosis of alpha-gal allergy, also known as mammalian meat allergy, was a complete surprise.

Alpha-gal allergy is a twenty-first-century allergy. Unlike respiratory, skin, and food allergy, alpha-gal was first discovered in the early 2000s, making it the perfect allergy to illustrate how allergy causation is a blend of our immune reactions, climate change, man-made ecological changes to our natural environments, and our lifestyles. It would take a team of scientists working on odd immune reactions to a new cancer drug to partially unravel it.

Enter Dr. Thomas Platts-Mills.

Platts-Mills is the director of the Division of Allergy and Clinical Immunology at the University of Virginia School of Medicine. We talked on the phone for more than an hour, in the midst of the second wave of the COVID pandemic. He's British, gregarious, and affable, and often halted our conversation to tell a joke or a story about one of his relatives (who are, quite honestly, an amazing bunch). For starters, he told me he doesn't like it when people refer to the allergy "epidemic." Using the word "epidemic" makes it seem to him as if rates for hay fever, asthma, eczema, and food allergy all rose dramatically at once. What's more interesting, and truer, is that hay fever rates increased first, then asthma began to spike in the 1960s and '70s, and then eczema and food allergy rates reared their ugly heads in the 1980s and '90s. More recently, we have begun to see more allergies that are not like their counterparts, in that they are not mediated by IgE antibody reactions. One is eosinophilic esophagitis (EOE), which we'll see again in the next chapter, and the other is alpha-gal.

Alpha-gal allergy is technically categorized as a food allergy, but it is an immune reaction to a sugar molecule found in most mammals (galactose-α-1,3-galactose, or alpha-gal) rather than to a protein, which is more typical. It's induced via a tick bite. The process is similar to how our immune systems can become sensitized to peanut proteins that leak through our skin barrier. When the tick bites us, its saliva permeates our skin's barrier. The tick's saliva is itself irritating to our cells, and typically causes the skin around the bite to itch, but the saliva may also contain traces of alpha-gal, especially if the tick's last blood meal was from a mammal like deer that produces alpha-gal. Once we've been bitten by the tick, our cells learn to associate alpha-gal (a harmless sugar molecule) with ticks (a harmful parasite). In some people, this combination produces a new allergy, one that is triggered by the ingestion of meat that contains the sugar.

Alpha-gal allergy is currently spreading in the United States because the ticks that primarily cause it to develop, the lone star ticks, are being pushed northward due to the expansion of fire ant territory (a predator of ticks), climate change, and various other ecological shifts. Lone star ticks have already been found in places as far north as southwestern Connecticut, Cape Cod, and Canada—much farther north than their normal range. (Though it's hard to say what is "normal" in light of the cascade effects of recent climate changes.)

The story of the discovery of alpha-gal is long and twisting. It's like a mystery with Tom Platts-Mills as its lead detective, who cracks the case.

"Well, it all started with cetuximab," Tom said. "Cetuximab is a monoclonal antibody used for treating cancer. And we knew that this monoclonal was causing a lot of reactions in Virginia, a full two years before it got marketed to the public."

In fact, as an interesting aside, the study of the drug was abruptly interrupted after Martha Stewart was sentenced to jail time for selling four thousand shares of stock in ImClone (the developer of cetuximab) before its stock price plummeted the very next day on news that the U.S. Food and Drug Administration (FDA) had decided not to allow cetuximab to be marketed. For a long time afterward, people lost interest in the drug. Then, slowly, studies began to resume. It was during this resurgence in interest that a cancer clinic in Arkansas reported a death from anaphylaxis after the first infusion of cetuximab. Several other patients also reported negative immune responses to the drug. That meant that the cancer patients were not *becoming* allergic to the drug, they were already allergic to it *beforehand*. The question was: How?

Due to his expertise in immunology, Tom was called in to investigate. He asked if he could get his hands on blood serum from the patients in the trial, specifically blood drawn from the patients before they had taken the experimental drug. The company running the trials, Bristol Myers Squibb, was keen to find out what was going wrong and helped Tom connect with oncologists from Vanderbilt University in Tennessee. Tom ended up receiving blood serum from approximately forty patients and forty matched controls (also from Tennessee). His research team measured each sample's antibody reactions and found that the patients who reacted poorly to cetuximab had an IgE antibody reaction to the drug molecule, indicating an allergic reaction. Patients without an antibody response did not experience a reaction to the drug.

Their interest piqued, and hot on the trail of a new allergic immune response, they tested another patient group from Texas. Only one patient had the antibody response. Perplexed, Tom and his group tested samples from patients enrolled in the trial from Boston. Nothing. That's when Tom realized that what they were seeing in the lab had nothing to do with having cancer, or taking the drug alone, and everything to do with living in central Tennessee.

"The patients had an IgE response to galactose-α-1,3-galactose, which is a mammalian oligosaccharide (or simple sugar) which humans don't have," Tom explained. "So, we have antibodies to this sugar, but we don't have the sugar itself."

His team wrote up their findings and published them in *The New England Journal of Medicine*.[53] Then, Tom and the second author of the original paper, Beloo Mirakhur, went to ImClone to meet with the biochemist responsible for glycosylation (a process that stabilizes a chemical molecule). Monoclonal antibodies used for this cancer drug are produced in a lab using cells. As Tom explained, "Ninety percent of monoclonals are made using a Chinese hamster ovary cell line, which doesn't produce alpha-gal." This means that most monoclonal drugs are perfectly safe to take even if a patient happens to have an allergy to alpha-gal. Cetuximab, on the other hand, was manufactured using a different type of cell—one that *did* make the sugar molecule alpha-gal. Now Tom and his team had both a smoking gun and an assay, or experimental method, to test for alpha-gal. Invigorated, Tom wanted to definitively prove that alpha-gal was causing the immune reactions.

"I told my team to get blood off of anybody who was standing still in the clinic and tell me who has this antibody!" Tom recalled, chuckling at his aha moment.

After testing everyone, Tom realized that some of his own patients had tested positive for reactivity to alpha-gal.

"They were all telling this stupid story that if they ate pork, they got hives four hours later," Tom said, remembering his initial dismissal of these patients. At first glance, the story seemed preposterous since most food allergies are fast-acting, manifesting approximately twenty minutes after ingestion. "If a child who is peanut allergic gets a peanut by accident in McDonald's, they'll know before they even leave the McDonald's," he said.

Tom also told me that it's not unusual for food allergists to hear stories about what their patients think is causing their reactions that cannot possibly be true. Often, the nocebo effect is at play. In other words, a person who believes that a food causes a negative reaction will often experience that reaction after ingesting it, even if that substance isn't actually at fault.[54] Before discovering alpha-gal reactions in the lab, delayed reactions in patients seemed highly unlikely. But from his team's initial findings, Tom realized that his patients were likely reporting a real allergic reaction, just one that no one had ever seen before.

Now Tom and his research team had a different puzzle to solve. How were people becoming sensitized to alpha-gal in the first place? What were the common denominators of all the cases?

First, every person who showed evidence of alpha-gal reactivity was located in just seven states: Virginia, North Carolina, Tennessee, Kentucky, Arkansas, Oklahoma, and southern Missouri. Looking at the cases plotted on a map, it looked like a long strip cutting across America. Tom assigned one of his lab technicians the task of figuring out what other things mapped onto these same geographic locations.

After a few days of googling, the lab tech came back and said the only thing he could find that matched the lab data was the CDC's map of cases of Rocky Mountain spotted fever, a tick-borne infectious disease. That was Tom's first clue that alpha-gal might also be tick-borne. He went back to all the patients who had positive alpha-gal reactions and discovered they all had one thing in common: They spent an inordinate amount of time outdoors.

"They were gardening, hiking, riding horses, hunting, you name it," Tom said.

This is when alpha-gal allergy gets really, really interesting from the perspective of climate change. As one of the top immunologists in the world, and as a respected historian of his own field of allergy, Tom argues that one of the reasons we haven't seen alpha-gal before now is that we've been altering our ecosystem over the past few decades.

Enter the chief scientist and state entomologist at Connecticut's Center for Vector Biology & Zoonotic Diseases, Dr. Kirby Stafford III.

Stafford has more than thirty years of experience tracking tick populations. When I spoke to him in late November 2021, it was one of the first truly cold days of the year, though average daily temperatures had yet to dip consistently into the low thirties on the Fahrenheit scale. I wanted to know more about how climate change, such as milder winter weather, and ecological shifts, such as an uptick in invasive plant species, were affecting the habits of ticks, especially the tick that causes alpha-gal.

In the United States, the tick that most often triggers meat allergy is the lone star tick. Lone star ticks are not particularly picky eaters. Though often found on deer and wild turkeys, they can also feed off mesomammals (which are medium-size mammals like racoons) and birds. But unlike the ticks that carry Lyme disease (those are black-legged ticks), they don't often feed on smaller rodents like field mice or chipmunks. Their typical geographic range, until quite recently, was the southern United States. Several biological and social factors have led to an expansion of their range into new northern territories, and there's also been a population explosion within their normal habitat.

But what is the biggest factor in the past few decades that has led to the proliferation of not only the lone star tick but all tick species? The explosion of its mammal hosts, in particular, at least for lone star ticks, the increase in the populations of white-tailed deer and wild turkeys.

"There's probably more white-tailed deer in the U.S. today than there were back before the colonists started wiping out the deer population," Stafford said.

Both ticks and deer were abundant in colonial America. In the mid-1770s, a Finnish naturalist named Pehr Kalm published a book about his travels in North America. In it, he mentions how bad the lone star ticks were, complaining that he could hardly sit down without a tick scrambling on him. But just a century later, the New York entomologist Coe Finch Austin noted in his own book that there were no ticks to be found in the same areas Kalm had reported being inundated with them. The reason? Likely the decimation of deer populations and the clearing of forests for farmland and fuel.

"In Connecticut in 1896," Stafford said, "it's estimated that there were only twelve deer left in the entire state. At that point, because of the low numbers, the state started regulating hunting to allow the population to increase."

Some New England states even imported deer from other locations to try to remedy the problem. They did the same, Stafford said, for wild turkeys and pigs. But today, states like Connecticut have the opposite problem—too many white-tailed deer. With dwindling hunters, less access to private land for the hunters who remain, an absence of natural predators like wolves and bears, and societal pushback to the idea of culling herds, the deer population has skyrocketed.

"We're also providing ideal suburban habitats for deer," Stafford said. "We're providing nice salad bars for them."

And that means the tick populations that they host—including lone star ticks—are also burgeoning. Stafford is seeing more and more lone star ticks submitted at the Connecticut state tick-testing laboratory. And while the percentage of lone star ticks is still dwarfed by that of black-legged ticks (coming in at just under 4 percent of the total in 2020), their numbers are growing significantly.

"So, we're not talking about big populations yet," Stafford said, "but the clock is ticking."

Get it?

But puns aside, Platts-Mills knows that more lone star ticks means more lone star tick bites. And there's something unique about the bite of the lone star tick: It tends to itch. The ticks that carry Lyme typically don't produce itching bites. In other words, the bite of the lone star tick induces a noticeable immune response.

"So, there's something in the saliva of this tick which induces a switch to IgE antibodies from a patient that's already got antibodies to the sugar," Tom explained. "We don't have this sugar naturally in our bodies, so we can make antibodies to it in the gut."

Even someone without meat allergy will produce antibodies to alpha-gal when they consume meat. But—and this is a big *but*—not everyone will make alpha-gal antibodies after a lone star tick bite. And even when someone makes the antibody, they won't necessarily have a negative immune response. Some will and some won't. Tom estimates that the presence of the antibody for alpha-gal in the general population is about one in five. But the disease only happens in a fraction of that population.

"And we don't understand that," Tom said.

In Australia, where they also have alpha-gal allergy, the same tick that causes alpha-gal can also cause anaphylaxis. But it doesn't cause anaphylactic responses here in the United States. Tom thinks the key to understanding why lies in understanding the composition of the saliva of the tick itself. It's the tick *plus* its saliva that is likely triggering the allergic response. In other words, adding a tick bite before the consumption of red meat is the magic ingredient that produces meat allergy.

When I asked him if there is a genetic component to alpha-gal, Tom told me that ticks like to bite some people more than others.

"If you put four people up on the Blue Ridge Mountains," he said, "two of them will be covered in ticks and the others won't. Why? Maybe they smell different. Does taking Lipitor [a popular cholesterol drug] change the smell of your skin? Does washing your skin change its smell?"

These subtle differences might matter to the lone star tick's palate—and that is something we don't fully understand. Could there be a genetic component to this? Certainly. But it could just as easily be due to the deodorant you slathered on that day, or the body wash you used in the shower. Or the drugs you're taking for your medical conditions, which are also partially driven by your genetics.

Tom posits that alpha-gal allergy might also be related to changes in the composition of bacteria that make up our gut microbiomes.

(Callback to Nagler's work; everything is connected!) There may be bacteria living in some people's guts that protect against the development of alpha-gal, even if they're bitten by a lone star tick and have antibodies to alpha-gal. Research has also shown that you're more likely to get the new tick-borne meat allergy if you have B-negative blood type, but no one knows why.[55] Finally, in the population of people with alpha-gal, 50 percent are allergic to pollen, dust mites, etc., and 50 percent are not allergic to anything.

"It's not the same genetic group as the people with hay fever or the people with food allergy," Tom said, returning to my original question after a long explanatory detour. "So it's not quite clear to me if this is genetic or not."

Like most of the allergic responses that came before it, the more we discover about alpha-gal, the more questions we have about its causes. In 2009, there were twenty-four reported cases of alpha-gal allergy; as of 2020, there were more than five thousand reported cases.[56] The true incidence rate of alpha-gal in the general population, however, remains unclear. So, too, does the risk of developing alpha-gal. As of this writing, scientists are unable to predict who will develop the condition after a tick bite, or even which ticks can cause the disease.

"A BETTER LIFE": AN ASIDE FOR PARENTS AND FUTURE PARENTS

At the end of our foray into how our modern lifestyles and our daily habits might be contributing to our allergy woes, I want us to pause to reflect again on Elizabeth and her children. Elizabeth's misplaced sense of guilt about being the indirect cause of her children's severe food allergies is intimately tied to her desire to provide them with the best possible care. Her children's suffering causes her suffering, something any parent or caregiver reading this can surely empathize with. Elizabeth's decision to allow her very ill babies to be treated with antibiotics at the emergency room as they struggled to stave off terrible infections was almost certainly the correct one, especially since infections, if left untreated, can have more dangerous outcomes. Yet her sense of regret lingers years later. After talking to the caregivers of many allergy patients, I'm certain that she is not alone.

Our efforts to make sure that our children have the best possible lives can—and often do—give rise to anxiety over the thousands of tiny choices we must make throughout their early childhoods. For many parents, that sense of unease begins even sooner, during pregnancy, as they research how best to prevent negative outcomes like severe allergies in the first place. Someone reading all the evidence I've collected here might be left feeling terrible about their necessary C-section or their reasonable choice not to breastfeed—even if those decisions were categorically the right ones to make given the circumstances.

In the information era, we're bombarded by advice from a variety of sources (ranging from legitimate evidence-based medical websites to dubious online YouTube videos) about what to do and what not to do to keep ourselves and our children healthy and happy. This entire book, in a sense, is no different. One could read any of the information here and use it to try to "game the immune system," so to speak. I don't recommend it. The best thing you can do is to follow the best available medical advice as immunologists learn more about how our immune systems react and respond to our new and ever-changing environments.

In other words, allergy parents and patients, give yourselves a much-needed break. None of us have the power to be the cause of our own allergies. The reality is far more complicated than that.

THERE ARE NO EASY ANSWERS TO ALLERGY CAUSATION, JUST DIFFICULT QUESTIONS

At the end of our exploration into the possible underlying causes of the dramatic surge in allergies over the past two centuries, what have we learned? We've discovered that while genetics plays a significant role in our immune system function, it does not adequately explain nor predict who will develop allergic conditions. We saw that while the environment we live in—the natural and man-made world around us—is most definitely contributing to the problem, it isn't the sole cause either. And our habits and behaviors matter greatly to our overall health and immune function, but they, too, cannot fully explain what is happening to our immune systems. What is happening to us is a result of all the things we've been doing differently over the past two hundred years and their effects on the environment and on our own biology. It's that simple and that complicated.

The eczema expert Peter Lio of Chicago would like for people to stop looking for a root cause for their conditions. He argues that it's the wrong question to be asking, especially since there probably isn't a simple root cause. But often, his patients don't want to hear the truth—that the cause of their symptoms is much more complicated than any one theory might suggest. But he tries to be honest with them anyway.

"I tell them it's a big mess," he said. "There's a skin barrier thing, an immune system thing, and there's something about the nerve endings, and then there is a behavioral piece. . . ."

The food allergist Pamela Guerrerio wanted me to underline that, ultimately, it is wrong for researchers to even look for a single cause. It sends an inaccurate message, she argued, that we just need to figure out what the problem is and then we will solve all of this. If the general public thinks that allergy is a simple problem, then they will just get more and more frustrated as the decades roll by without a straightforward solution. And, as we'll see in part 3 on treatments, most of the solutions we have for allergies are imperfect at best.

"My message would be that there isn't one cause," Guerrerio said, "and that we need to understand that there is probably a genetic susceptibility, and on top of that is environmental exposures, and what's true in one group to account for their increase in allergy may be different for another group. There are just so many different changes in environmental factors."

At Cincinnati Children's Hospital, in a discussion about pollution's effect on asthma, Dr. Neeru Khurana Hershey summed it up best:

"There's no one thing. If there were one thing, we would've found it and we would've figured it out. It's a combination of things and it's different for different geographic areas and different genetic backgrounds of people. Sometimes it's easier to put the blame on something than it is to take a hard look at what we're doing and how we're contributing to the problem. Because everyone's contributing to the problem. What are we doing as a society that we should be doing better? And what am I as an individual doing that contributes to that? These are not easy questions or answers. These are hard ones."

PART THREE

Treatments

o o o

Since the discovery of hay fever in 1819, physicians and allergy sufferers alike have been searching for medical treatments that might alleviate symptoms or cure allergies outright. But as we've just seen, if the causes of the malady are not simply biological, then the remedies we seek are likely to go beyond basic pharmaceutical solutions. In Part 3, we'll examine all our efforts—past and present—to cope with the burgeoning problem of allergies. From the big business of allergy drugs to the social and governmental policies enacted to help deal with a growing set of environmental problems, we'll see that the solutions to our irritations are just as complex as their causes.

CHAPTER 7

Remedies for the Irritated: Allergy Treatments Past, Present, and Future

---

THE TREATMENT TALE OF THE UNSTOPPABLE EMILY BROWN: ADVENTURES IN FOOD AVOIDANCE

"Food allergy was not even really on my radar," Emily Brown tells me.

Growing up, she had hay fever and asthma, but that was it. She still has environmental allergies. Her husband has them, too. But no one in their family had ever had any food allergies whatsoever.

"I taught preschool before I had my kids, and I did have a couple of kids in my class that had food allergies," Emily says. "But that was really my only exposure. It didn't show up in my personal life until I had kids."

In 2011, Emily's older daughter was born. She had severe eczema from birth. "She didn't sleep well," Emily explains, "and was very colicky. She was just not a happy baby."

As a new family, Emily and her husband were trying to cope with their daughter's discomfort. Then, when her daughter was six months old, she developed blood in her stool. Emily brought her to the pediatrician. Just wait and see, she recalls the doctor telling her.

Back in 2011, guidelines for pregnant women and new parents were different from what they are now. Parents before 2016 were advised to avoid introducing common allergens like peanuts or strawberries into their infants' diets. Emily still consumed them during her pregnancy and while breastfeeding but decided to heed medical advice and wait a full year before introducing her older daughter to peanuts.

"I gave her peanut butter for the first time and her face just immediately began to swell," Emily recalls. "She developed hives. It was very scary. I didn't know what to do."

Emily immediately called her pediatrician, who told her to give the baby some Benadryl. When her daughter went in for allergy testing, they discovered she had sensitivity to peanuts, milk, eggs, wheat, and soy. Emily remembers being given a prescription for an EpiPen and told to avoid all those foods. That's not an easy thing to do when the list of things to avoid is so long.

"I remember leaving and thinking, Well, what are we going to eat now? How do I cook?"

Emily's mom was just as perplexed. So was Emily's best friend, an obstetrician and gynecologist in her third year of residency. As a medical student, Emily's friend had only been given an hour's lecture on the topic of allergy in medical school. Quickly realizing that she was going to have to help herself, Emily started doing research online. First, she cleaned up her kitchen and read all the labels on the food she already had, throwing out anything with the allergens her daughter was sensitive to. Her mother gave her a gift card to Whole Foods to help her restock.

Emily had taken a year off to be with her infant daughter, to have that special experience of bonding and learning to be a mom. The plan had always been to go back to work, but the diagnosis of food allergy made that impossible. The preschool she worked for wouldn't allow Emily to bring her daughter in because she had "too many" allergies. And if her daughter couldn't come to work with her, then Emily couldn't afford to pay for childcare to go back to work herself. The irony of a preschool teacher needing to find a preschool that would take her daughter isn't lost on me. While talking to the parents of children with severe allergy, I have often wondered how they find the wherewithal to cope with the effects of the diagnosis on their collective lives.

"I always tell people it was kind of like this perfect storm," says Emily. "Our expenses increased significantly. Our grocery bill quadrupled overnight. I had this perceived loss of income because even though we didn't have it yet, we had been planning for it. My husband is a social worker and hadn't had a raise in six years. So things were really tight, but we were making it. I mean, it was really, really tight."

Emily pauses and takes a breath.

"Then that spring, my mother passed away," she says, her voice quieter but firmer. "I haven't talked a lot about it publicly. The only reason I bring it up is because sometimes I think people just assume that when you go through difficult times that you'll just have family or somebody to help you, right? But you don't always have that. My mother passed away, and for me, I lost a big part of my familiar support."

Things got even more complicated for Emily and her family when she found out she was pregnant with her second child. When that happened, Emily's husband suggested that they enroll their family in the federal Women, Infants, and Children program (WIC) because with an additional child, it would be hard to feed everyone on his salary alone. Emily had grown up in a solidly middle-class family. No one had ever been on any kind of federal assistance. But she also knew that her husband, in his job as a social worker, counseled his clients to sign up for WIC all the time.

Emily remembers him telling her, "We've paid into the system. Just enroll. It'll just be temporary."

So, Emily enrolled.

In our conversation, Emily emphasizes how grateful she was for the support from WIC, and what a terrific program it is. But at the same time, its restrictions on which brands and sizes of food were eligible hampered Emily's ability to find food for her allergic child. The federal assistance program was simply not set up to deal with families contending with a severe food allergy diagnosis, whose only real treatment option was avoidance of the offending allergens.

"We were the poster children for what this program was supposed to support, but we could not fully participate in it because the foods that we needed weren't there," Emily explains. "There were a lot of challenges. You could only get tortillas that are thirty-two ounces. And for whatever reason the store might only have the sixteen-ounce tortillas in the eligible brand. In that case, you can't get it. Literally you have to leave it on the shelf. You can't make any substitutes."

In Emily's case, her daughter couldn't eat any products that contained wheat, but the standard approved bread on the WIC program is whole wheat bread. Not to mention that wheat, as Emily is quick to point out to me, is in practically every processed food available on the grocery shelf.

To me, this situation sounds like a complete nightmare. Emily tells me that it was even worse than that. When you're on the food assistance program, you're restricted to buying the food in your county. So, if a supermarket in the next county over has what you need, you still cannot buy it. The products available in the same store chain might be completely different depending on which county you're shopping in.

"I happen to live in the poorest county," Emily explains, "and we're right next to the richest county. So, when I talk about access to food, there's just so many layers."

When Emily called the state WIC office to explain her situation, she was told to get a doctor's note in order to put her daughter on a special infant formula. But when she called her pediatrician, the doctor explained that her daughter should be eating solid foods, not formula, at her age. Then, when she went back to WIC with what her doctor had told her, the WIC office told her to call her local food bank because they couldn't help her. But when she called them, the food bank said the same thing, because the way they packed food exposed everything to flour dust, which contains wheat gluten. The large food bank recommended a smaller local pantry to Emily, since local pantries tend to rescue food from grocers like Whole Foods. Eventually, Emily found a local pantry that opened at one o'clock. She got there a half hour early, but the line was already wrapped around the building. She discovered that people often lined up three hours ahead to get the best selection of available foods. That first day, Emily ended up standing in line for hours.

"You know, you watch the carts roll by. And when you're hungry, you're just so hopeful that there's going to be something there for you," she says, reliving that day over again, the same disappointment and disillusionment clear in her voice. "But by the time I got in to shop, there were literally only two potatoes and a jar of salsa. Those were the only safe foods in that pantry for my family."

That moment crystallized for Emily a much larger problem for low-income people with food allergies, whose only available treatment option is food avoidance.

"When food avoidance is the prevailing standard of care," Emily argues, "then access to safe food is not just a matter of sustenance. It's a matter of treatment and therefore it should be covered. There should be a way to ensure that all patients have access." Emily said that part of the problem is the power imbalance inherent in the American social and healthcare systems. "There's also this prevailing idea that beggars can't be choosy, right? There's no dignity there."

For Emily, the availability of allergen-free food is about providing safety nets for *everyone*. Her family had never been on assistance before, and only needed it because of her child's diagnosis, and the hope was that even that would be temporary. But the safety net simply wasn't there for her or her family. So Emily started a nonprofit dedicated to fixing this problem. The Food Equality Initiative serves about 150 families in Kansas City, Missouri, but Emily hopes to expand it to serve other areas. Emily said that her motivation for starting her nonprofit was the knowledge that she could not be the only one dealing with this problem.

"I knew there were other families, but I knew that they couldn't speak up for whatever reason because this whole experience, it doesn't make anyone feel good about themselves."

There is, Emily explains, so much shame involved at every step of the process. She wanted to change that, and she knew she was good at doing research and standing up for others, so she dug in. She became determined to help not just herself and her own children with allergy but everyone else like them. So in 2015, Emily borrowed $500 to start a food pantry for people with food allergies. It was the first allergy-friendly and gluten-free food pantry in the entire United States.

"I always joke with people that the pantry is my third child," she says, laughing, "because it took nine months from the idea to when we actually opened."

She partnered with pantries and negotiated shelf space, then stocked those shelves herself and trained local food pantry staff and volunteers to help food allergy families shop. Emily estimates that there are between eight thousand and fifteen thousand people who would qualify for the service in her area, but she can only afford to serve 150 families—or 1 percent of the estimated need. Money is the main problem, but Emily also wants to make sure that the Food Equality Initiative focuses more on values than on volume. She's happy serving smaller numbers of people if she can focus more on serving them well.

"While we do want to serve more people," she says, "we want to make sure that the service we're providing is meaningful, that it adds value, and that it's done in a way that provides dignity."

Emily ensures that no one ever stands in line at the Food Equality Initiative. Her services are appointment based, so families don't have to waste time to get food—time that could be spent looking for a job or doing something else.

Emily is very vocal about the glaring disparities in the diagnosis and treatment for allergies that run along racial and economic lines. Black and Hispanic children are far less likely to be diagnosed with food allergy and far more likely to end up in the ER with anaphylaxis. They are also less likely to have access to good healthcare and new treatments like oral immunotherapy (OIT) for peanut allergy (more on that in chapter 9).

"What's mind-boggling to me is that food allergy was really thought to be an affluent, very white disease," Emily says, "and you hear allergists say, 'Well, I just don't see any patients of color.' I think that's where we start to see the role of implicit bias. The doctor's not doing anything malicious, but when they see a Black patient or a

Puerto Rican patient come in, they're not asking the right questions. There's a real disparity in diagnostics."

As both an allergy mom and the director of a food allergy non-profit, Emily argues that we need to offer better information to patients in emergency rooms and better follow-up care.

"There's just such a lack of knowledge in some of these communities and it's not because they're inept or don't want to be compliant. It's because they simply haven't had access. It's a real challenge."

It's the same trouble with allergy treatments. While Emily is excited about all the new allergy drugs and treatments, when she looks at the studies, she's disturbed to see that the overwhelming majority of trial participants are white. She specifically references the drug Palforzia, a newly FDA-approved peanut allergy treatment, in which 90 percent of study participants were white.

"When I talk to patients of color about OIT, they've never heard of it. To me, it seems like a treatment that patients of color didn't ask for, don't know about, weren't included in developing—just completely left out. It feels like a solution for the wealthy."

She's worried, quite legitimately, that only white patients will receive OIT, while Black patients (and everyone else) will be left with food avoidance as their only treatment option. She wonders when, if ever, we will design solutions based on *every* patient's needs and not just on those of white or wealthy patients. Emily sees her entire life now as a concerted pursuit of health equity in allergy. She knows how difficult it can be for people of color and poor patients to get treatment because she has lived experience.

Emily's husband was laid off from his social work job not too long after their daughter was diagnosed with allergy. He found a new position working at a call center to support his growing family, but it didn't offer health insurance. Emily's family had to go on Medicaid—so they also lost their access to allergy specialists.

"None of the private allergists in our community accept Medicaid," she explains. "The only place you can go is the academic centers, which have tremendous waits. I remember trying to get in and it took us six months to get an appointment."

At the same time that Emily was starting her nonprofit, her younger daughter became very, very sick. By the time she turned one, she had been diagnosed with failure to thrive and put on feeding therapy. Around the age of four, she was diagnosed with a rare form of food allergy called eosinophilic esophagitis (EoE). People with EoE have an accumulation of eosinophils, a type of white blood cell, in the lining of their esophagus. These cells react negatively to food allergens and can cause irritation, pain, and constriction of the esophagus, making it hard to ingest food. EoE is a rare disease, affecting approximately one in two thousand people. Luckily, there is an EoE clinic in Kansas City, where Emily's daughter received care. To help diagnose her, doctors at the clinic prescribed a series of elimination diets: first for milk, then a four-food elimination, then an eight-food elimination, then eight plus beef and chicken.

In 2019, Emily attended an allergy conference as an invited speaker (the only way, she said, that she could afford to go). While there, Emily talked to a top EoE researcher who recommended that she try a treatment that involves a regimen of oral steroids in addition to a limited diet. Emily did and her daughter went into remission. She's seven now and still in remission, even though they don't know all her triggers. (As we've already seen, sometimes that's impossible to figure out in the best-case scenarios.) The next step, without the steroids, would have been a supplemental nutritional formula, which isn't covered by insurance in Missouri. Emily and her husband would have had to consider moving across the state line, where it is covered, to avoid the cost of $3,000 per month. The steroids have worked, but the side effects are a constant concern. Her daughter's cortisol and endocrine levels have to be closely monitored, in case the steroids are damaging her organs, and the dosage might need to be adjusted if her levels change. If that happens, she might fall out of remission. Children don't usually grow out of EoE; it's a lifelong disorder.

———

Emily Brown's allergy treatment story is both unique and yet all too familiar. It highlights the trials and travails that allergy sufferers and their families go through after their initial diagnosis. As we've seen, the process of allergy diagnosis is complicated, and the process of allergy treatment can be even more so. Each patient's symptomatology is different because each patient's biological response is distinctive. This can make the treatment of allergy extremely difficult.

Adding to that is the fact that allergy treatments did not progress for the larger part of the past two centuries. Until very recently, we've been stuck with the same options for treatment. All have benefits and drawbacks, and none provide full, lasting relief. In the rest of this chapter, we'll survey the history of treatments for allergy patients— from morphine and asthma cigarettes in the nineteenth century to Xolair and Aimmune Therapeutic's Palforzia today. We'll look at what has changed, and what hasn't, and the difficulties allergy patients face when it comes to relieving their worst symptoms. Finally, we will see that there are no easy "cures" for our irritated immune systems, though hope for better prevention and relief might be on the distant horizon.

THE MORE THINGS CHANGE, THE MORE TREATMENTS STAY THE SAME: THE PAST AND PRESENT OF ALLERGY CARE

In an 1868 letter to his physician friend Oliver Wendell Holmes, the famous abolitionist and clergyman Henry Ward Beecher complained of his inability to find any remedies for his hay fever. Holmes replied: "Gravel is an effectual remedy. It should be taken about eight feet deep." Holmes's friendly joking over Beecher's despair would have come out of a deep frustration with the prevailing allergy treatments of the day. Even physicians of the era who had allergy themselves, such as Dr. John Bostock and Dr. Charles Harrison Blackley, could not find adequate relief for their worst symptoms, despite decades of self-experimentation and clinical practice. As much as our scientific understanding of immune functions has deepened over the last century, many of us are familiar with Beecher's woes. Allergy treatments have, until quite recently, remained largely the same.

To better illustrate how the medical approach to allergies has shifted throughout the last two centuries, we will follow three typical patients undergoing a variety of treatments for some of the most common allergic reactions. Each case presented here is a fictional composite based on all the patients I have interviewed, interacted with, read about in historical texts, or heard about from clinicians. I have crafted them as white, middle- to upper-middle-class persons with good healthcare coverage who reside in cities or suburban areas and have relatively easy access to allergists. That's because, at least historically speaking, most allergy patients were white, well-off, and urban—and as of this writing, those demographic groups are still most likely to have the best access to the best treatments. We'll continue to interrogate the serious and persistent social problem with allergy care later in the chapter.

Respiratory Allergies and Asthma

Jennifer is a young, otherwise healthy woman in her mid- to late twenties who has had seasonal and environmental respiratory allergies since she was a child. She's particularly affected by oak, grass, and ragweed pollens. She also suffers from moderate to severe asthma, which is fairly well controlled unless the pollen count climbs too high, or Jennifer exerts herself too much, especially when outdoors during the summer months. As someone who has played sports like soccer and softball her entire life, avoiding outdoor activities isn't really an option for Jennifer. On a few occasions, a particularly severe attack has sent her to the hospital.

The 1800s through the 1930s: If someone like Jennifer had lived in this time period, she likely would have had access to a few different treatment options, though none of them would have been particularly effective. To begin with, her family physician would likely have counseled her to try to avoid the dirty or polluted city air that was suspected of contributing to her symptoms. So, too, any trees or flowers or odors or substances like dust that seemed to lead to attacks of asthma. Put simply: Remove all known allergens from her vicinity.

Patients with respiratory allergy were often instructed to take down drapes and curtains and take up rugs; remove all pictures and other "dust catchers" for thorough cleaning; go over all furniture, even bedsprings, with an oiled or damp cloth; wash floors and radiators; and get rid of animals in the home. Jennifer may also have been counseled to seal up all her windows and stay indoors as much as possible throughout the warmer growing seasons.

If her family was particularly wealthy, Jennifer's doctor may have recommended that she pack up her household and abscond to the mountains or the seashore during the summer months. Beginning in the late 1800s, it was common for wealthier patients to travel to a "health resort," typically located either high up in the mountains or in the desert, to escape from city air and pollen. In the eastern half of the United States, large, luxurious hotels sprang up in the Adirondacks and the White Mountains to cater to hay fever, asthma, and bronchitis sufferers. In the American West, the mountains of Colorado and the Arizona desert became popular destinations.

"The rate of asthma sufferers in Tucson today is higher than in other parts of the nation," the historian Gregg Mitman explained to me, "because you had this influx of asthma immigrants, who not only increased the genetic load in the population, but also brought in plants like mulberry and olive trees to green the desert that ended up exacerbating their allergies." (The consequences of planting these trees are discussed in chapter 10.)

If Jennifer's family was not rich enough to relocate nor amenable to the idea, then she could avail herself of a variety of drugs that were prescribed either to prevent her asthma attacks or to control her worst symptoms during bouts of lung and nasal irritation. In a text written in 1934 for medical practitioners, Dr. Samuel Feinberg recommended the following as possible treatments: asthma powders and cigarettes, morphine and opium derivatives, iodides, ether, cocaine, alcoholic drinks, and calcium.[1] Adrenaline was also prescribed for sudden, acute attacks of asthma.

Another standard option for controlling respiratory allergy was injection with pollen extracts. Desensitization, as it was called, was first introduced in London in 1911 in order to "build up immunity" to an allergen, curtailing attacks before they began.[2] As a candidate for this treatment, Jennifer would have had three methods to choose from: (1) Her physician would introduce trace amounts of an allergen via small scratches on her skin; this was called the cutaneous method. (2) Her physician would give her the allergen intradermally, as an injection, with controlled dosing. In this case, the doctor would need to craft their own formulation containing the local allergens that were plaguing Jennifer. (3) Her physician would place a small drop of a solution containing a measured amount of the allergen directly into the conjunctival sac of her eye, producing small ophthalmic reactions. Over time, and through repeated administration of small doses whose potency increased incrementally, Jennifer's immune system might be trained to tolerate the offending allergens. But desensitization was not successful in every case (data on the overall success rate of these early treatments is largely nonexistent). If none of these various treatments was effective, then Jennifer may have been given more dubious injections. It was not uncommon at the time for physicians to try to prime a patient's immune system using other substances, including milk, calcium, sulfur, turpentine, and small amounts of tubercle bacillus. Clinicians hoped that injections of such substances might induce tolerance and help relieve symptoms. Patients might also be given injections of solutions of their own blood serum or microorganisms collected from their own respiratory tract or of "parasitic extract," specifically *Ascaris lumbricoides,* also known as large roundworm, the most common human parasite.[3] One famous early allergist noted that some physicians were experimenting with injections of flu and typhoid vaccines, bacteria from the intestines, and snake venom. He complained about a lack of empiricism in much of the experimentation with these allergy treatments, arguing that most were nothing more than a "by-product of worthless dross."[4]

If nothing else worked for Jennifer, she may have been subjected to surgery to attempt to correct any respiratory tract abnormalities. By the 1930s, it was a common practice to remove the tonsils and adenoids of patients with respiratory troubles, though there is scant evidence that it worked. Jennifer may also have been taught a series of special breathing exercises, combined with new postures (a straight back and dropped shoulders), and given regular massages to encourage the development of a "mobile chest wall."[5]

The 1940s through the present day: As the decades wore on and medical technology evolved, allergists experimented with allergy-proof chambers, air filtering masks, and nasal douches with a special apparatus to spray CO_2 into the nasal cavity;[6] electrotherapy, actinotherapy (or the use of UV rays), and X-ray therapy. Depending on her physician, Jennifer might have undergone any of these speculative therapies to try to control her seasonal attacks.

But the typical treatments for respiratory allergies and asthma stayed more or less the same for decades. In 2022, the standard treatment for patients such as Jennifer, with allergic diseases due to extrinsic allergens, remains avoidance. Avoidance is the preferred treatment, because the removal of the allergen from the patient's environment or the patient from the allergic environment means that the immune system will not be exposed to any triggers and will not have to react in the first place. If, for instance, the offending allergen is cat dander, this may be doable (though not emotionally simple): One can give away the cat, clean the house of dander, and avoid cats in the future. But this method obviously isn't doable for those of us who, like Jennifer, have sensitivities to things that are more ubiquitous such as tree or grass pollen and dust mites, which can be difficult, if not impossible, to remove entirely from our homes.

The second line of defense against allergic disease—again, largely unchanged since the nineteenth century—is thorough cleaning and readjustment of the lived environment. For instance, if someone is allergic to rodents, dust mites, cockroaches, or molds, then regular cleaning of the household, washing of all bedding, and removal of the pest's sources of food, water, and entry to the home can cut down on exposure to the offending allergens. Air conditioners and air filtering devices are often recommended to patients since these newer technologies can filter some or most of the allergens like pollen from the air. Seasonal allergy patients like Jennifer are also told to shower as soon as they get home, to wash pollen off their skin and out of their hair, and to change their clothes immediately. All the effort required to avoid coming into direct contact with allergens can feel onerous and exhausting to patients with moderate to severe allergies. And most of the time, despite our best efforts, total avoidance is not feasible.

Jennifer's third line of defense is attempting to control her worst symptoms by using one or more pharmaceutical drugs. Many of the drugs that would be prescribed to Jennifer today, in the early 2020s, would have been familiar to the Jennifer of the 1940s (or even much earlier than that).

"We're still using drugs that have been used for a century," Dr. Robert Schleimer at Northwestern University explained to me. "You know, humans haven't changed that much, and the drugs haven't changed that much. We just keep adding new or better versions."

One of the first lines of treatment for mild to moderate respiratory allergy is—and has been for nearly a century—antihistamines. Antihistamines are the oldest class of drugs still prescribed to allergy sufferers. They were accidentally discovered in 1937 during research on immune function in guinea pigs; by 1942, antihistamines had become widely available for general use and were quickly popular as a *somewhat* effective remedy. Antihistamines work by binding onto the immune cell's histamine receptors instead of histamine, thus blocking the development of a histamine reaction. As of this writing, there are ten first- and second-generation antihistamines approved in the United States that someone like Jennifer may take to control her runny nose and itchy eyes, available either over the counter or with a prescription.

During our discussion of modern treatments for respiratory allergy, Schleimer noted that second-generation antihistamines such as Zyrtec and Allegra tend to have milder side effects than first-generation drugs. Although they remain effective, first-generation antihistamines have sedating effects and can—and often do—interfere with mental focus and processing (as anyone who has ever taken Benadryl as an impromptu sleep aid can attest). The newer drugs are non-sedating and avoid most of the worst side effects of the first-generation drugs.

So, why keep manufacturing first-generation antihistamines? Why would anyone still prescribe or use them if their side effects are so dramatic? The short answer is that first-generation drugs are more drying and thus more effective in combating the nasal congestion that often goes along with mild to moderate respiratory allergies. The second-generation drugs, although they work just as well to halt the histamine reaction, don't do anything to help with congestion, so some patients like Jennifer actually prefer taking the older, harsher drugs during the height of pollen season.

I want to pause briefly here to note that patients like Jennifer often develop extraordinarily strong attachments to and beliefs about specific brands or types of antihistamines. In interviews with allergy sufferers, I often hear detailed explanations of drug regimens, and tales of how one drug was working fine for years and then suddenly stopped working, or that a new antihistamine had changed someone's life, or that they had built up a tolerance to a particular brand of antihistamine. All of that being said and acknowledged, no scientific studies have shown that patients actually can, or do, develop a physical tolerance to any of the ten approved antihistamine drugs. Researchers who work on the topic think that allergy patients stop taking their prescriptions, or want to switch their prescriptions, because of the persistence of severe symptoms while taking the drug, rather than the fact that the drugs have become less effective over time. It's possible, too, that the patients' symptoms are worse due to a larger pollen load or more exposure to higher levels of allergens. In other words, we feel like the drugs we're taking aren't working because sometimes they aren't.[7]

"Modern pharmacology focuses more on what's called an unmet need," Schleimer said. "The unmet need is all the people with severe disease that the commonly effective drugs are not working on."

He gave me a pertinent example. Seventy years ago a lot of people like Jennifer were suffering greatly or even dying from asthma. For many, available drugs were not effectively controlling their disease. Then oral steroids were developed and approved for general use. But those oral steroids turned out to be highly dangerous to the asthma patients for whom they were prescribed: Steroid use can cause serious damage to bodily tissues over time.

"But that was better than dying of asthma," Schleimer said, shrugging his shoulders. "Those same drugs are still around, by the way, but the death rate from asthma has plummeted because of the development of inhaled steroids that were an improvement on those oral steroids. And inhaled steroids work locally; they don't cause all the horrible side effects that the oral steroids do. So sometimes it's taking the same drugs and making them better, improving them. For the antihistamines, there are multiple generations now. Each one improved. The same thing with the steroids."

Treatment for asthma mirrors that for seasonal allergies but adds on new medications to control for more severe symptoms. You have probably seen someone using an inhaler to control their breathing or to help prevent an attack. Those inhalers contain different medications, depending on the needs of the patient.

Typically, respiratory symptoms in asthma patients are controlled using a category of pharmaceuticals known as beta-agonists. Beta-agonists are bronchodilators used for short-term therapy to control symptoms. Six are approved for use in the United States. Some are short-acting inhalers like albuterol, levalbuterol, and pirbuterol; others are more long-acting versions like salmeterol, formoterol, and vilanterol. Basically, these inhalers are our modern way of dosing out ephedrine, a drug prescribed by doctors a hundred years ago to improve the lungs' breathing capacity.

Long-acting inhalers are prescribed for daily use only in order to control chronic or persistent asthma. Physicians generally do not recommend their daily use over time, since decreases in lung capacity and increased reactivity in the lungs have been found after a duration of as short as three weeks of continuous daily usage.[8] Short-acting inhalers are usually prescribed for temporary control of symptoms or to prevent symptoms from occurring before strenuous activity or exercise—like Jennifer's soccer games. The side effects of persistent use of beta-agonist inhalers can include body tremors, a feeling of restlessness, and stomach troubles.

Inhaled corticosteroids (commonly referred to as steroids) are a first-line treatment for all patients with persistent asthma who are not responding to beta-agonists. Currently, there are eight inhaled corticosteroid preparations approved for use in the United States. Once control of her worst symptoms is achieved, a patient like Jennifer is placed back on the lowest dose of steroids possible to control her disease. Intranasal glucocorticoids, a different class of corticosteroid, are used for allergic rhinitis and are considered safe and effective because they deliver local and targeted treatment. These sprays, including Flonase, Nasonex, and Nasacort, are often recommended as a first-line therapy for perennial and seasonal allergic rhinitis.

There is a brand-new class of biologic drugs developed for asthma, monoclonal antibodies, but they're expensive. (If this sounds familiar, it's likely because monoclonal antibodies were being used to treat COVID-19 patients.) Dupilumab, marketed as Dupixent, is a new and highly effective antibody treatment for asthma and eczema. It currently costs an average of $36,000 per year. Those prices might be worth it if something could truly cure allergy or completely erase symptoms, but, as Schleimer reminded me, these new classes of drugs don't come without side effects either. Approximately 25 percent of patients on dupilumab go on to develop conjunctivitis. It can also cause elevations of eosinophils, or a type of white blood cell found in our body tissues. And those elevated levels of immune cells, as we saw with Emily Brown's younger daughter, can spell bigger trouble like EoE.

"So that drug is not going to be the panacea," Schleimer said. "But for now, it's the best drug that we have."

"Think about it: Thirty years ago we basically only had antihistamines available, and it was quite obvious to all of us that they barely work," Dr. Alkis Togias at the NIH told me, as we sat in his office discussing available allergy treatments. "They barely work, but that was all we had for medication. Until recently, the challenge was in the lack of treatments. Today that's less of an issue. We have some quite sophisticated treatments, but some of them cost thirty thousand dollars a year—so how can we implement that on a large scale? Physicians have a brand-new problem. As a doctor, you should not care about anything but giving the best possible treatment. But physicians are being bombarded by cost-effectiveness."

In sum, Jennifer's treatment options now are better than they have ever been. Ultimately, she will do a bit of everything to try to manage her condition. She may decide not to play soccer on a particularly bad day for grass pollen. She might buy herself an air purifier and take a daily antihistamine or use a beta-agonist inhaler from March to October each year. During the summer months, when the grass pollen is high, she might carry a stronger steroid inhaler to the playing field to ward off an attack or help to control one that's been triggered. All of this together might make Jennifer feel better most of the time, but she—and others like her—likely still struggle to control their worst symptoms. And all of this, of course, also depends on what her insurance will cover, what her co-pays will be, and how much or how little she can afford.

Food Allergy

David's parents first noticed a problem when they tried giving him a small amount of peanut butter at six months: He broke into a serious rash. After having their young son tested, David's parents discovered that he was allergic to peanuts, tree nuts, and eggs. Over the years, David has ended up in the emergency room a handful of times for an accidental exposure, frightening both him and his parents. Now ten years old, he is careful to avoid accidentally ingesting any of his trigger allergens, but sometimes he feels anxious when he attends playdates or birthday parties.

The 1800s through the 1930s: If David had been born during this time period, it's far less likely that he would have developed food allergies at all. Children who did experience negative reactions to food were all given the same treatment regimens. In an early medical text on "alimentary anaphylaxis," or gastrointestinal food allergy, specialists warned that overfeeding infants and children was a direct cause of anaphylaxis and should therefore be avoided. To help control the reactions, all sensitizing foods were to be withdrawn from the child's diet. Finally, in serious cases where avoidance alone was not sufficient, shock treatments were recommended, as well as all-liquid diets and "injections of camphorated oil, ether, and adrenalin."[9] (Not unexpectedly, the shock treatments were not effective.)

By the 1930s, however, the standard recommendation for treatment of someone with David's allergies would have been an elimination diet. Dr. Albert Rowe, a renowned food allergy expert, believed (largely erroneously) that elimination diets would induce desensitization in patients as they gradually added back in foods to which they were sensitive. Dr. Rowe also experimented with "protein therapy," wherein patients were injected with either milk or tubercle bacillus (similarly used to treat respiratory allergy patients) but found the results in his patients "disappointing." With an eye toward preventing the development of food allergy in infants, Dr. Rowe gave the following advice to pregnant mothers: (1) no overindulgence of foods during pregnancy; (2) no overfeeding in infancy; (3) no intermittent feeding of specific foods to their infants—their diets should be consistent; (4) no early feeding of food known to be allergenic in others (as we've already seen, this last piece of advice would eventually prove to be especially disastrous).[10]

In his own book on allergy, Dr. Arthur Coca recommended that physicians design individualized treatments based on the particular causes of food allergy in patients. Prophylaxis, or preventing allergy, should always be step one in any treatment plan.[11] If a patient had an attack, then the first choice to subdue symptoms should be epinephrine, followed by ephedrine. Other treatments were rest, catharsis (or purging), starvation for twenty-four to forty-eight hours, counter-irritants (applying cups or poultices to irritate the skin), and shielding patients from extreme temperatures and drafts of wind.

But for the most part, beyond avoidance of the allergenic food, David and his parents would have had few treatment options that worked. Little was known about food allergy at the time, as relatively few patients developed the allergy in the first place. It would take many more decades for research on food allergy to improve treatments and care.

The 1940s through the present day: Food allergists today are still very limited in the treatments they can offer, despite the growing number of patients worldwide. Avoidance remains the standard treatment protocol. This advice usually includes the removal of all foods containing the allergen from the home or, if complete avoidance isn't feasible, vigilance against cross-contamination of cooking utensils, plates, or pans. Today, David's parents would likely first change their own eating habits to make sure that David was not accidentally exposed to eggs or nuts from food cooked in their own home. When going out, they would try to ensure that none of the allergens were in the food he ordered or consumed—which is often trickier than one might assume, especially at social gatherings. The effort to avoid the allergen, and avoid an anaphylactic event, can be exhausting.

"People are desperate," said Dr. Pamela Guerrerio, a food allergist at the NIH. "There is no treatment for food allergy. I have nothing to offer you other than an EpiPen."

In addition to making sure David avoided all his allergens, his parents would likely also carry an EpiPen to administer if David accidentally ate something that contained one of his triggers. An EpiPen is the most famous brand of auto-injector for adrenaline. Immediate injection with adrenaline can delay the biological processes that lead to anaphylaxis and, in severe cases, death. Using adrenaline usually gives patients like David enough extra time to get to the hospital for lifesaving treatment.

Yet a recent study found that 52 percent of adults with life-threatening allergies didn't use EpiPens at all.[12] Of those who had been prescribed an auto-injector, only 89 percent had filled their prescriptions. Study participants who didn't fill the prescriptions in the first place cited the high cost of the injectors and a lack of prior severe reactions as reasons. Even when people did fill their prescriptions for an EpiPen, around 21 percent of those surveyed said they didn't know how to use it. Another 45 percent said they didn't have their auto-injector available the moment they experienced a severe reaction, rendering the prescription effectively useless.

Adrenaline injectors themselves are not small and need to be stored within a specific temperature range (and so, for example, cannot conveniently be stored in a vehicle glove compartment on a hot summer day). Many allergy patients have privately told me that carrying their auto-injectors at all times is not always practical or preferable. Teenagers and young adults, specifically, argued that the need to carry one at parties and other gatherings is often embarrassing as well as inconvenient. Parents with younger children like David, though, almost always carry EpiPens and most schools have them on hand for emergencies.

Outside of avoidance and auto-injections of adrenaline in case of accidental exposure, the only other therapy available to food allergy patients like David is immunotherapy. Immunotherapy is defined as the process by which a small amount of an allergen is administered to the patient over time in order to build up immune tolerance. It's usually only recommended when a person has a severe form of allergy and other therapies such as antihistamines and avoidance are not effectively controlling their worst symptoms. The treatment process can take years and its effectiveness varies greatly from person to person (this is especially true in relationship to respiratory allergy). We don't know exactly why immunotherapy works, but the process can indeed alter our underlying immune mechanisms. As a therapy for both food allergy and respiratory allergy, immunotherapy shows promise not only in preventing symptoms but in curtailing the number of anaphylactic events a patient like David will experience in their lifetime.

There are three basic types of immunotherapy: subcutaneous (SCIT), sublingual (SLIT), and oral (OIT). SCIT and SLIT are used for environmental allergies like Jennifer's. OIT is used to treat food allergy. For all three types of immunotherapy, a patient usually undergoes treatment in a clinical setting. It is important to have directly observed immunotherapy, in the rare case that the patient develops a severe response to the minute quantity of allergen administered during treatment. Patients like Jennifer receiving SLIT are given either drops or tablets. Tablets are placed under the tongue for a few minutes, until they begin to dissolve, and then swallowed; drops are placed under the tongue for a specified amount of time, then swallowed or spit out. Patients receiving SCIT get shots weekly or biweekly in the beginning of therapy. Doses continue for three years, after which most patients can safely stop treatment and maintain its protective effects. The positive effects of immunotherapy can last for years after the treatment is discontinued—but they won't last forever. Patients like David receiving OIT will ingest low, measured doses of their food allergen under clinical supervision. The initial appointments take hours because the patient needs to be monitored for reactions after consuming their allergen. Then, for two weeks, the patient continues to take a preset amount of allergen daily at home. David would need to return to the pediatric allergy clinic every two weeks to increase his dosage of allergen by a preset amount and then be monitored for any reaction. This part of the treatment process is known as "up-dosing" and continues for several months. At the end of the up-dosing phase, David would then need to continue to consume a certain amount of his allergen to maintain his level of desensitization. (If OIT still seems confusing at this point, don't worry. We'll get more into the nitty-gritty of how new OIT treatments work in chapter 9.)

Possible side effects of SLIT, SCIT, and OIT are swelling of the tongue, mouth, or throat, itching of the mouth, and—in rare cases—anaphylaxis. Severe swelling at the site of shots during SCIT is rare but can occur. All three forms of immunotherapy are generally considered safe treatments, but effectiveness varies.[13] Until recently, the gold standard treatment for most allergy has been SCIT. In past studies, SCIT had been found to be more effective, even though SLIT was considered a safer treatment (in short, this means it caused fewer anaphylactic events). Unlike drugs that help with symptoms, SCIT, SLIT, and OIT treatments work to actively change the patient's overall immune functioning, helping immune cells learn to tolerate greater amounts of the antigens before responding.

Interestingly, and somewhat problematically, only some of the allergen extracts used in immunotherapy are standardized using FDA regulations and calibrated according to standard potency levels called biologic allergy units (BAU). Other extracts used in immunotherapy are standardized by measuring the specific amount of allergen contained in the extract (usually measured in weight per volume).

In other words, and as we've seen so many times before, there isn't much global standardization. Each manufacturer chooses which standard to follow for crafting their extracts. Until recently, that also meant that not all immunotherapy was equal. Two allergists working in the same metro area might use different extracts from different manufacturers. The patients' responses, too, will vary.

And for many, immunotherapy simply won't work. This is especially true if a patient has environmental allergies like hay fever (which is why I didn't even list SCIT or SLIT as an option for Jennifer's oak, grass, and ragweed allergies). If a patient hasn't seen any improvement after a full year of undergoing immunotherapy, it's recommended that the treatment be discontinued. Researchers are currently working on the development of simple blood tests that might be able to predict whether or not immunotherapy is likely to be effective before a patient is offered the treatment in the first place.[14] Such a screening test would save patients time, money, inconvenience, and the emotional distress that often goes along with undergoing such an extensive therapy.

It's important to note that only a handful of the allergy sufferers I spoke to in the course of researching this book had tried immunotherapy. Of those who had, a few found it to be beneficial. Most had discontinued the therapy after a few months to a year, telling me that they couldn't discern much benefit in the way of symptom relief and were tired of going in to see their allergist on a weekly or biweekly basis. None of the adult food allergy patients I spoke with had tried immunotherapy for their allergies; some mentioned their fears about ingesting the very thing that might kill them in a larger dose. OIT treatment also causes reactions such as upset stomach, mouth tingling, and rash. Some patients are not willing to undergo treatment that, at least in the first few months, causes them so much discomfort. Many young children find the treatment itself to be anxiety provoking, since the symptoms are similar to ones they've had before, at the start of an anaphylactic event.

There is a new FDA-approved, standardized preparation for oral immunotherapy for peanut allergy, available since January 2020, called Palforzia. Today, David's parents would have the option to choose Palforzia as a treatment to significantly increase David's tolerance to accidental exposure to peanuts. It has been shown to be incredibly effective, allowing patients to consume a few peanuts without a severe immune response. For David and his parents, that might mean a lot less anxiety.

But as we'll see in chapter 9, an immunotherapy like Palforzia is not a "cure" for food allergy. A food allergist I interviewed cautioned me more than once that Palforzia, while helpful, was only effective for a single allergen: peanut. Most allergic children are reactive to multiple allergens. Immunotherapy for food allergy also has to be maintained indefinitely or its effectiveness wanes over time. In other words, immunotherapy is not a panacea for food allergy. Dr. Kari Nadeau stressed the need for experts to push themselves to do better, arguing that a single new FDA-approved drug is not nearly enough: "We should not just sit and applaud ourselves for any given therapy right now for food allergy since there are still safety issues."

Ulitmately, David's parents would have to make a decision between the traditional course of action—food avoidance plus an auto-injector of adrenaline for emergencies—and immunotherapy. For now, those are their only choices. And neither of them is a perfect option.

Atopic Dermatitis, or Eczema

Emma is six and loves dogs and cats, but she is wildly allergic to them. She's also had severe eczema since she was an infant. Her skin flares can last for weeks to months, and she has trouble sleeping and concentrating in school. Typically, the skin on her cheeks, elbows, knees, and hands becomes inflamed. Emma's skin turns itchy and bright red; her scratching often leads to open sores. Sometimes her outbreak is so severe that pus forms on the skin of her hands.

The 1800s through the 1930s: Historically, people with Emma's condition would have been categorized as simply having a skin eruption or rash. "Eczema" was coined in the mid-1700s but wasn't considered to be an allergic condition until the 1930s, when the term "atopic dermatitis" was first used. Until the modern era, the typical treatment for the condition would have been the application of various poultices (pastes made of natural materials, similar to our modern skin mask treatments) or, in some cases, bloodletting. As understandings of the immune system progressed, many early physicians started to link eczema to a patient's diet—in particular, to the consumption of milk. Dietary restrictions to control outbreaks would have been common. But Emma's family would not have had many options or effective treatments to choose from, leaving Emma to suffer through on her own, hopefully to outgrow the condition.

The 1940s through the present day: Until recently, most of the treatment options for eczema have been terrible. From my perspective, after talking to dozens of allergists and many people with the condition, eczema is the most difficult of all the allergic conditions to treat. For starters, as Dr. Peter Lio pointed out to me, most patients with atopic dermatitis have no idea what triggers their outbreaks. It could be environmental allergens; it could be chemicals in something they are using; it could be something as simple as heat or exercise. Avoidance, then, is typically not an option.

And the symptoms of atopic dermatitis can be unbearable. For decades, patients trying to manage their condition have been in a lose-lose situation.

"It's on your skin, so everyone can see it at all times," Dr. Jessie Felton told me during a Zoom chat about her experience as a pediatric dermatologist at Brighton and Sussex University Hospitals and the Royal Alexandra Children's Hospital in the United Kingdom. "It's on display, particularly if you have it on your face. It affects the comfort of your skin at all times, so it literally affects everything you do. It affects your concentration. It affects sleep. And that's not just sleep deprivation for just the child; it's the whole family."

When I ask her about treating atopic dermatitis in patients like Emma, she tells me the story of a similar patient, a five-year-old girl with severe eczema. The case was so agonizing that the child was placed on immunosuppressants—a last line of defense.

"The child had told her mum to cover every mirror in the house because she couldn't bear to look at herself," Felton said. "And that's a five-year-old. She was just so distressed by her skin—bright red, broken, dry peeling."

Typically, mild cases of eczema are initially treated with routine daily skin care. Patients are asked to apply special moisturizers to the affected areas several times per day, although recent studies have suggested that there's less evidence that moisturizing is preventative. In more persistent mild or moderate cases, topical corticosteroids are prescribed as a first-line treatment. There are currently seven different groups of topical steroids, ranked from the most potent (betamethasone dipropionate) to the least potent (hydrocortisone) and prescribed accordingly. It's not uncommon for eczema patients to develop bacterial infections as well, so antibiotic ointments and oral antibiotics may also be prescribed to help get a skin outbreak under control. Management of any food allergies or respiratory allergies the patient has may be recommended, since they might cause eczema to flare up. Second-line defenses include systemic corticosteroids (taken either orally or given by injection) and immunosuppressants. Both of these more severe options work, but they also have rebound effects once the treatment is discontinued. In other words, successfully treating eczema can actually make eczema worse. Immunosuppressants have an additional nasty side effect: Because they work by partially turning off general immune response, they might actually increase the risk of developing infections.

Felton said that treatment of eczema is difficult for patients like Emma not only because the treatments are harsh and often don't work, but also because of the mental toll it takes on everyone involved. "Moms and dads feel so guilty when they're applying the creams," she explained. "It's very uncomfortable for the child. There are battles every night. So the parents are trying to comply with treatment. At the same time, the child's screaming at them not to do it, and it's really, really hard."

Treatments are also difficult because each patient is different. And typically, by the time they've made it into Felton's office, they've weathered a lot.

"By the time I see patients, they've been through everything," she said. "They've probably had private allergy tests done. They might have gone abroad. They are buying all kinds of creams that they would have spent quite a lot of money on and been very frustrated that nothing has helped."

The first thing Dr. Felton does is to try to figure out what is going on with the individual in front of her. She describes each case of atopic dermatitis as a ball of yarn, all tangled up. It takes patience to unravel it. She sees hope, too, in several new treatments that are in development, including anti-IgE monoclonal antibodies and Janus kinase inhibitors (we'll take a more in-depth look at these drugs in chapter 9). The monoclonal antibody Dupixent was the first approved biologic drug for atopic dermatitis. Felton argues that it is a game-changer, the first new treatment to show real promise for the most severe eczema cases, but there are still all sorts of hoops to jump through for approval to prescribe it through the National Health Service in the United Kingdom. (I heard similar complaints about the difficulty in getting some American health insurance companies to cover the cost.)

Still, even with the hassles of getting approval to use it, Felton is excited to have a new option. With steroids, you have to do regular blood tests to make sure the treatment isn't affecting the patient's liver or kidney functions. Still, Felton uses them in her treatment of skin conditions, since she has found they can be highly effective when taken for several months (often clearing the skin without a rebound effect). One of the most popular oral steroids, prednisone, doesn't require blood monitoring and can provide much-needed relief, but it isn't a good long-term option for patients (due to side effects like higher blood pressure). Topical steroids reduce inflammation but can also increase blood vessel formation, thin the skin, and cause immune complications. Topical calcineurin inhibitors (TCIs) have fewer side effects than steroids and can help reduce itch, inflammation, and dryness, but long-term use may lead to an increased risk of cancer (as studies have demonstrated in animals). With the new biologics[15] in development, there will be less worry. And, if you can control the skin's reaction for a year, most children like Emma will not rebound once the treatment is discontinued.

"There is something really magical about children's immune systems," Felton said, "and that doesn't happen in adults."

But some patients are still failing even on Dupixent. Dr. Emma Guttman-Yassky at the Icahn School of Medicine at Mount Sinai cautioned in an article published by the National Eczema Association that none of these newer drugs will "cure" atopic dermatitis. (And while it's true that many children "grow out" of their eczema as adults, their underlying atopy remains and may flare up at a future point—say, when their immune system function is compromised by age or stress.) And as with the older treatments, there's potential that these newer drugs might be effective at clearing a patient's skin for several years, until the immune system adapts to them. After that, the patient might experience outbreaks again and need to switch things up, leading to the necessity for ever-newer, better drug treatments.

This cycle of the "unmet need" will continue until we learn enough about the underlying biological mechanisms of all allergic immune responses to prevent them from occurring in the first place. Until then, patients like Emma, David, and Jennifer are stuck in a game of cat and mouse, trapped between their irritated immune systems and the treatments we have available to calm them back down. The more we know about how allergies work, the better off we'll be at preventing and treating them.

TOMORROW'S TREATMENTS? THE PROMISE OF
NEW TECHNOLOGIES ON THE HORIZON

One of the things that both allergists and allergy patients can agree on is that the currently available treatments are not nearly good enough. Although we've made progress, we're still mostly using things we discovered over a hundred years (or more) ago. However, the development of new scientific technologies is helping us to start to turn the corner on allergy treatments, perhaps paving a new path forward. What all the innovations coming online in the next few decades will have in common is machine learning and new laboratory techniques.

The magnitude of computing power that scientists have access to today is incredible and only accelerating, leading to the development of more sophisticated algorithms that help immunologists sort through the enormous amount of patient data they've been amassing for decades. Some of those efforts are already showing promise.

Using computer algorithms, researchers in Sweden found two new biomarkers (a few genes) that may help dermatologists distinguish between irritant eczema and contact dermatitis (an irritation of the skin caused by a substance like poison ivy or fragrances).[16] Because it can be very difficult to tell the difference between these two immune reactions based on visible symptoms alone, misdiagnosis is common. There are also two different types of eczema: allergic contact eczema and nonallergic irritant eczema (caused by chemicals or events like exercise). New diagnostic tests developed using these biomarkers might additionally help to distinguish between types of eczema and lead to better treatments.

The biotech company AllerGenis, which licensed rights to the research in allergenic epitope (the part of a protein that antibodies attach to) analyses done by the renowned food allergist Dr. Hugh Sampson, uses immunological research to develop new technologies for allergy diagnosis and treatment. AllerGenis is currently using machine learning (a type of artificial intelligence) to predict the outcomes of milk allergy immunotherapy with 87 percent accuracy (which is higher than the rate of currently available and more common blood serum tests). It uses data from immunoassays measuring an individual's antibody responses to predict the effectiveness of immunotherapy.

Trying to harness the power of big data, Anthem Blue Cross Blue Shield (one of America's largest health insurance providers) teamed up with researchers at Harvard University to see if artificial intelligence can improve patient treatment outcomes for those with allergies.[17] The groundbreaking collaboration hopes to use the insurance company's treasure trove of patient data to figure out which treatments are more effective—and for whom. Such research might enable more personalized treatment protocols and eliminate the need for trial and error during the early stages of allergy care . . . and would also save health insurance companies a lot of money in the process.

Other researchers are trying to apply the power of new laboratory techniques and decades of prior research in immunology to find new biological mechanisms to target for treatment. Researchers at Aarhus University in Denmark recently discovered an antibody that may block IgE from attaching to cells, preventing the release of histamine.[18] Scientists at La Jolla Institute for Immunology are attempting to block the accumulation of harmful T cells in the lungs during an asthma attack by blocking signaling proteins.[19] And research at Northwestern University found that inhibitors for an enzyme found in mast cells called Bruton's tyrosine kinase (BTK) can impede allergic reactions from escalating, possibly preventing anaphylaxis from occurring.[20] The study used three different BTK inhibitors[21] to block the release of histamine and other allergenic signals in mast cells in a test tube. This finding could open up the possibility of developing a drug that would prevent anaphylactic responses in people with life-threatening allergies.

Meanwhile, the National Institutes of Health is using the collective power of academic, governmental, and industrial research labs to discover new allergic pathways and new approaches to treatment. Dr. Marshall Plaut, an internationally recognized immunologist who recently retired from the NIH, told me that some of the most promising research in relationship to the future of allergy treatments is happening in the Immune Tolerance Network (an NIH-funded collaboration). "It has a model that proposes an allergen+ method, or an allergen in combination with some synergistic molecule that could hasten tolerance induction," Plaut explained. By which he meant training the immune system to tolerate an allergen using another molecule to speed up that process, perhaps something that the immune system already recognizes and doesn't negatively react to. The goal of the Immune Tolerance Network is to treat allergic diseases *before* they appear.

In that same vein, a lot of new scientific research is focused on deploying immunotherapy techniques, or vaccines, to help with allergy. BM32, a new vaccine under development for grass pollen and currently in clinical phase 2 trials, may reduce the symptoms of respiratory allergy caused by pollen by up to 25 percent.[22] The treatment requires far fewer doses than traditional forms of immunotherapy and has fewer side effects. Another vaccine under development in Switzerland, HypoCat, works by immunizing cats against their own major allergen (called Fel d 1). The vaccine produces antibodies that bind to Fel d 1 to reduce secretions. That means that cats who have been treated with the vaccine will produce lower levels of allergens, alleviating cat dander allergies. The company is also working on a vaccine for dogs based on similar principles. Flinders University in Australia is testing a new bee venom vaccine that uses advax adjuvant (the same adjuvant used to improve flu vaccine efficacy) to dramatically speed up immunotherapy.[23] The current vaccine for bee venom allergy necessitates fifty injections over a three-year period. And new research on an old form of immunotherapy is unearthing new evidence that eczema might be treated with allergy shots (or immunotherapy shots), which would expand treatment options for atopic dermatitis patients.[24]

Taking the underlying idea of immunotherapy to new levels, researchers at Duke University are using nanoparticles to "retrain" the immune system to tolerate food allergens.[25] The nanoparticles are loaded up with cytokines (a group of small proteins used in cell signaling) and antigens and then delivered into the patient's skin. Once introduced into the body, they travel to the lymph nodes, where they can provide protection against anaphylaxis by introducing antigens to our immune cells in a friendlier manner. Relying on a similar principle, cutting-edge research being done at Northwestern University suggests that a nanoparticle containing gluten can help train the immune system of celiac disease patients to tolerate it.[26] (Important note: Celiac disease is not an allergy. It is an autoimmune response triggered by the ingestion of wheat.) The patient's macrophages (large white blood cells that can surround and kill other cells) take up the particle and present it back to the immune system in a way that increases immune tolerance. In a clinical trial, Northwestern's nanoparticle treatment showed a 90 percent drop in the levels of immune inflammation in patients who were exposed to gluten. In other words, nanotechnology is showing great promise as a new way to induce greater tolerance. The hope is that, once fully developed, the treatment might be used to prevent allergic disease in those individuals with a genetic predisposition.

Another possible "cure" for allergic disease may lie in good, old-fashioned (relatively speaking, at any rate) genetic technology. Using the latest genetic techniques for cell manipulation, University of Queensland researchers have been able to wipe out the memory of T cells in animals, allowing their immune systems to tolerate sensitizing allergenic proteins.[27] The study used an asthma allergen, but scientists think that the same principle might apply to other classes of allergens, such as bee venom or shellfish. The ultimate goal of this line of scientific inquiry is to be able to develop a single-injection gene therapy that would alter the decision-making process of a patient's T cells, promoting tolerance to any number of allergens.

On the flip side, genetic researchers are attempting to design genetically modified organisms (GMOs) that lack the proteins responsible for causing most food allergy in the first place. Dr. Eliot Herman at the University of Arizona is working to produce a soybean that doesn't make the protein that triggers an immune response in people allergic to soy.[28] The soybean is crossed with a more common line and produces very low amounts of the offending protein. Herman is currently testing it on pigs bred to have an increased sensitivity to soy (which some pigs already naturally have, as some of us do). Pending results, we may have a new tool in our fight against food allergy soon: less antigenic foods.

But while all of this scientific research is promising, we have to remember that the process that leads from scientific research to a new treatment is incredibly slow, expensive, and arduous.

"We're now actually able to take about fifty years of immunology research and apply it to really help patients," Dr. Marc Rothenberg at Cincinnati Children's Hospital said, "but we're not yet able to cure allergies."

That being said, Rothenberg told me that he's certain that research being performed today will help pave the way toward a future cure. Fundamental scientific research on the biological mechanisms behind allergy—or what we often call "basic science"—is the key to scientific progress. As we've seen throughout the history sections of this book, most of the advancements in our understanding of immunology have been made by "accident," or when smart people were able to follow their instincts or natural curiosity. Rothenberg argues for the free and open sharing of research to advance the objective of a cure for allergy, and not just as a means to an end to apply that knowledge as a Band-Aid to treat its symptoms. In fact, he is notorious for releasing his own data to the public via a website that he has set up called EGIDExpress (see https://egidexpress.research.cchmc.org).

"There's too much red tape, too much regulation," Rothenberg explained. "Even though we have a great team, we're limited by time, and we're limited by money. It takes thirty years from the time of discovery until new drugs are approved by the FDA. That's a long time to wait when you have kids suffering from these diseases."

This was something I heard from almost every scientific researcher working on allergies. They need more money so they can hire more people and buy more tech and do more research. Without more investment in basic immunological science, better allergy treatments and a "cure" remain decades away.

A NOTE ON ALTERNATIVE MEDICINE, THE PLACEBO EFFECT, AND THE QUEST FOR RELIEF

I'm in a dermatologist's office in midtown Manhattan, getting my annual cancer screening, when I tell her about this book. She's relieved that someone is writing an evidence-based book because she knows that patients often do their own research—usually via Google and WebMD—and try to self-diagnose and self-treat their skin conditions. It's hard enough, she tells me, to treat these patients without also needing to battle the sometimes false or misleading information that they discover online or from friends. Most irritating for her, however, is the fact that her own sister-in-law won't listen to her—a board-certified dermatologist with years of experience—about her niece's eczema. Her sister-in-law listens to someone on YouTube who lives in South Africa and has just purchased a supplement that my dermatologist knows has zero chance of helping her niece.

"It's really, really frustrating," my dermatologist says, using a loupe to scan the freckles and moles scattered around my skin. "I understand that she's desperate to help her kid. I just wish she would listen to me. To the science."

I struggled with how to talk about alternative treatments here, because while I don't want to give too much oxygen to theories that have no scientific evidence to back them, I also don't want people who have tried them to feel unseen or shamed. I, too, understand why, when biomedicine has failed to alleviate someone's nastiest symptoms, they turn elsewhere for help. Allergies are, at their worst, scary and miserable and exhausting. Treatments are as much about hope and belief as they are about science. The placebo effect is real: If you believe something will work, it might have a visible, measurable effect on your condition. So, can we say that something like cannabis or acupuncture or going to a salt cave doesn't work if a patient gets some relief from it?

"People look for anything that will help," Pamela Guerrerio told me when I visited her at the NIH, "whether it's Chinese herbal medicines or whatever."

None of the allergists or research scientists I spoke to condemned patients for trying to find solutions to their problems. But nearly all of them also expressed the same frustration as my dermatologist. Experts also underscored that not all alternative or complementary treatments are created alike: Chinese herbal medicine,[29] homeopathy, and acupuncture are a few that may have actual benefits and are currently being studied in clinical trials at places like Mount Sinai Hospital in New York City. Taking probiotics,[30] Reiki massage, chiropractic treatments, and consuming local honey have shown little to no benefit in controlled studies.

Many of the people with allergies who were interviewed for this book mentioned that they had tried one or more alternative therapies. Some reported that their symptoms improved. Some told me that those treatments, too, seemed to work for a bit and then failed them. Most patients used a combination of information sources to make treatment decisions, including their general practitioners, allergists, pharmacists, friends with similar allergies, family members, and sometimes strangers in online support groups. As a medical anthropologist, I would argue that alternative treatments can, will, and probably should play a role in treating allergy—ideally in tandem with biomedicine—since they offer patients not only potential relief (either through active substances or through the placebo effect) but, perhaps more important, hope.

According to many of the experts I talked with, the best approach is integrative therapy, or using a combination of methods and treatments to care for allergy patients. Dr. Meenu Singh at the Postgraduate Institute of Medical Education and Research in Chandigarh, India, believes that yoga helps her patients and often prescribes it to them in addition to more traditional biomedical treatments. She finds that by developing a yoga practice, her patients often learn to control their breathing—and thus to better control their asthma. Singh told me that the development of new biologic treatments, which are ten times as expensive as the cheaper, older treatments, makes no difference for her poor patients. But those cheaper treatments, like inhalers, present their own challenges. The parents and grandparents of Singh's patients are always worried that their children will become addicted to their inhalers or steroids. When the children start to get better, they start avoiding the medicine.

Singh noted, too, that allergy diagnostic tests are expensive; so are chest X-rays and CT scans. Allergy tests can cost ten thousand rupees for a whole panel, and general practitioners order them but then don't know how to interpret them. A good allergy history, she said, takes about an hour, and most doctors aren't willing to spend that time— or they can't because their clinics are jam-packed with patients. Time listening to your patients, Singh suggested, is the best medicine of all.

"Many times if they talk, they feel better," Singh told me. "And then they listen to you later on. They entrust you with their kids' health and so we really need to listen to them."

And that, in the end, might be the most effective alternative treatment for allergy: simply taking the time to listen to patients and witnessing the lived experience of their conditions. That might be what draws people to complementary practitioners and therapies in the first place. Peter Lio, who runs an integrative eczema center in Chicago, told me that he tries to spend as much time with patients as possible, and works with them to try out different therapies and treatments. Fundamentally, Singh and Lio know that what patients really want on their quest for allergy relief is to be heard.

ON ETHNICITY, SOCIAL CLASS, GEOGRAPHY, AND ACCESS TO QUALITY CARE

When Emily Brown's daughter was first diagnosed with food allergy, she scoured the internet for information and support groups. Eventually, she discovered a local group that met in the suburbs of Kansas City, Kansas. Emily would drive forty-five minutes once a month to attend. The meetings were usually held in a shopping mall at Panera Bread.

"I could never afford to eat anything," Emily explained. "We were struggling at the time, and the gas to get there and back was an extra expense. This was my only outing. It was like the highlight of my entire month."

The other moms in the support group had children of varied ages, so Emily found it extremely helpful to hear advice from people who had already gone through what she was currently experiencing.

"I felt like there was so much value there, but it was still awkward and hard. I was the only person of color in that group, certainly the poorest, and I remember they were all talking about their allergist," Emily recalled. "I was not going to an allergist. We were only going to our pediatrician, because even though we had insurance at the time, I couldn't afford the specialty co-pay. It only cost me twenty-five dollars to go to the pediatrician, but it costs me fifty dollars to see an allergist. And when you can't afford food, you're thinking about every little penny going out the door."

Basic allergy care and support groups are not as accessible as people often assume they are. Access to healthcare and affordable treatments depends, at least in the United States, on a whole host of factors, including the color of a patient's skin, their geographic location, and their economic status. It's not as simple as getting a diagnosis and then getting treatment.

"This is a disease that disproportionately impacts patients of color," Emily said, "but when you look at the patient advocacy world, you don't see patients of color. And that should always be a sign that something is wrong."

Emily argues that the plight of low-income food allergy families has been largely invisible, even to food allergy nonprofit organizations that focus on education and support, like Food Allergy Research & Education (FARE). And the same can be said for the plight of low-income asthma and eczema families, I can assure you. The common denominators in getting access to allergy treatment are disposable income and health insurance.

For the first few years after her daughter's food allergy diagnosis, Emily's family was on food assistance for nine months and Medicaid for a year and a half. She reminds me that everyone is vulnerable to a setback, and no one should think that they'll never need public services. It's in all of our interests, Emily argues, to make sure those services are as robust and responsive to all of our needs as they can be. She also worries about all the private donations going to fund scientific research instead of programs like her own nonprofit.

"If all the money is going to research for treatments, then we won't have good individual outcomes for families in need," she said.

The historian Gregg Mitman echoed Emily's point in our conversation. He questioned our historical focus on developing better drugs versus doing something about our environments. In other words, we can decide to take $10 million and try to get kids who live in the inner city earlier access to inhaled steroids, or we could invest that same amount of money to develop interventions that focus on environmental aspects of health—like where the bus depots, with their higher diesel emissions, are located in relationship to low-income housing. What Gregg and Emily are really asking is: Do all families have equal access to healthcare, early diagnosis, and healthy environments?

"It's hard to say where our money is best spent," Gregg argued, "because we've never done that comparative study. Everyone assumes that changes to infrastructure are more costly than providing access to pharmaceuticals."

But what if we're wrong about how we approach allergy care? What if the prevention of allergies is ultimately much cheaper than their treatment? And, more pertinently, even if we do develop more miracle drugs like Dupixent, who will have access to them? Will poor people, or lower-caste people, or people of color, or rural people who live in areas without allergy specialists, or citizens of developing nations have access to these newer, more effective treatments?

Alkis Togias returns to the question of healthcare cost when we are discussing allergy treatment and care. He argues that doctors should not carry the burden of worrying about the cost of the care they are providing. Physicians should only be worried about which treatments are the most effective. But in the absence of universal healthcare and subsidized drug costs, that isn't realistic.

"Practicing physicians are being torn down by the system, by insurance companies and all that comes along with this system, so it creates a big conflict," Togias says. "To me, that is another major challenge of allergy care."

As allergy rates worsen and our immune systems become more and more irritated, the market for allergy treatment continues to expand. In fact, the rise in allergy rates over the last century has created endless profits for the businesses that cater to allergy patients. Let's take a look next at the complex role that money currently plays in the treatment of allergic diseases worldwide.

CHAPTER 8

The Booming Business of
Allergy Treatments

———

Like every other chronic illness, allergy is very big business. The already staggering and steadily growing number of global allergy sufferers has meant that various industries, from pharmaceutical companies to food manufacturers to cosmetics companies, stand to make enormous profits from the creation of new or better diagnostic tools, allergy pills, inhalers, epinephrine auto-injectors, allergen-free product lines, allergen-free foods, and other products and services that cater to the millions of allergy patients worldwide. To quote the medical historian Mark Jackson, "At the turn of the millennium, allergy means money." How much money? A lot. Let me illustrate by giving you a quick rundown of the latest projections:

- The global sales of allergy diagnostics (tests) and therapeutics (treatments) combined is expected to climb to US$52 billion per year by 2026. For comparison, that is equivalent to Tanzania's yearly gross domestic product (GDP).
- China is expected to spend $8.7 billion on allergy care by 2027.
- In 2020, during the COVID-19 crisis, the global market for anti-allergy drugs was estimated at $24.8 billion per year, with $6.7 billion of that being spent in the United States alone. Market analysts anticipate that global number will rise to $35–39 billion by 2027—a growth rate of around 6.8 percent annually.

- The market for allergen-free foods is expected to reach $108 billion per year by 2030.

These are overwhelmingly big numbers. The rest of this chapter recounts three different stories about the rocky relationship between biomedical science and business. All three examine how money, specifically the quest for profit in a capitalistic healthcare system, affects the development and availability of allergy treatments. First, we'll dive into a worst-case scenario by looking at the 2016 EpiPen pricing scandal. Second, we'll move on to a best-case scenario by looking at the recently FDA-approved biologic drug dupilumab. Third, we'll learn how government funding seeds academic research on basic allergy mechanisms that then fuel various venture capital investments in new biotechnology start-ups that promise innovative allergy diagnostics or therapeutics.

TALE #1: THE EPIPEN SCANDAL

If you don't know what an EpiPen is, consider yourself lucky. The EpiPen is a ubiquitous presence in the lives of many millions of allergy patients, particularly those with moderate to severe food allergies or anyone else likely to experience anaphylaxis as a result of accidental exposure to an antigen. The EpiPen is a patented, six-inch-long auto-injector device filled with epinephrine (also known as adrenaline). There are other brands of epinephrine auto-injectors available, but EpiPen remains the go-to prescription for emergency use. Since it came on the market in 1987, the EpiPen has become synonymous with lifesaving injections of adrenaline. Much like using the name "Kleenex" in place of the word "tissue," when allergy patients talk to me about their prescription auto-injectors, they talk explicitly about their "Epis" or "EpiPens." In the United States, when someone needs a shot of adrenaline, what they call out for is an EpiPen. That's largely because EpiPen was the first auto-injector and continues to have a reputation for being easy to use. Over decades, people with allergies came to depend on an EpiPen as defense against the worst-case scenario: death via anaphylaxis. Doctors and patients trust them as a brand because they are reliable, and because they have helped to save many lives.

But there's another factor at play: EpiPen has been heavily marketed. When Mylan Pharmaceuticals acquired the company that manufactured EpiPen in 2007, it began heavily marketing its new product in public awareness campaigns about the very real and growing dangers of anaphylaxis. At the time, food allergy rates were skyrocketing, and the news was filled with tragic stories of the early deaths of small children and teenagers who had eaten a cookie containing peanuts or been given something with eggs in it by accident. The increase in anaphylaxis events gave Mylan an opportunity to promote its lifesaving product. As just one example of this subtle advertising campaign, in 2014 Mylan partnered with Walt Disney Parks and Resorts to craft a website and several children's books aimed at families with severe allergies.

In addition to increasing EpiPen's marketing budget, Mylan also began hiring a lot of lobbyists. According to the Center for Public Integrity, during the ten-year period between 2006 and 2016, Mylan Pharmaceuticals hired more lobbyists than any other U.S. company. (There are 1,587 registered pharmaceutical lobbyists currently working in Washington, D.C.) Their combined efforts paid off. The FDA changed the labeling of the drug to include people who were merely "at risk" for an anaphylactic event, rather than just those who had already experienced one.

Mylan's efforts also included lobbying in thirty-six states between 2010 and 2014. Why spend so much money courting state legislators when you already have the FDA on board? What was Mylan after? The answer: State regulations that required epinephrine to be made available in all public schools.

Legislation to require access to auto-injectors in school settings *was* direly needed, since quickly giving someone one or two doses of epinephrine can save their life in the event of a severe allergic reaction. However, such laws are also potentially very lucrative for pharmaceutical companies like Mylan that produce the devices. Each state that enacted such legislation meant thousands of schools suddenly in need of auto-injectors. In 2012, Mylan established the EpiPen4Schools program, offering free starter kits to participating schools (four free auto-injectors) and a discount on future EpiPen purchases.[1] The donated EpiPens were tax-deductible and a huge boost to the brand name recognition of EpiPen.

All of Mylan's marketing and lobbying efforts were extremely effective, if ethically odorous. Mylan's market share of emergency adrenaline auto-injectors went from 90 percent in 2007 to 95 percent—a near monopoly—by 2016.

As an interesting and pertinent aside, while researching this book, I learned new information about my father's death. According to my father's girlfriend, my father knew about his allergy to bee venom, but had never had anything like an anaphylactic response to a sting before his death in 1996. His doctor had given him a prescription for an EpiPen in the early 1990s, even though it would be "off-label" since my father had never had an anaphylactic reaction before (at the time, only people who had already had at least one severe allergic reaction would typically be prescribed one), and recommended that my father carry one with him during bee and wasp season "just in case." But because he did not have any history of severe allergic reactions, my father's insurance would not cover the cost of the prescription. My dad, being a very practical New Englander, did his own math and calculated that his risk was lower than the out-of-pocket cost of an EpiPen and the inconvenience of carting it around with him during the summer months. In the end, he quite literally paid for this decision with his life. If he had been diagnosed with a bee venom allergy a mere decade later, he would have been able to get his prescription covered via the FDA's new addition to the drug's labeling—potentially saving his life.

In the summer of 2016, Mylan Pharmaceuticals raised the price of EpiPen to just over $600 for a standard two-pack prescription. The patented auto-injection pens, filled with precise, lifesaving doses of adrenaline, were being prescribed each year to around 3.6 million

American patients with severe or life-threatening allergies. Don't worry, I'll do that math for you: At that price, EpiPen's manufacturer pulled in around $2.16 billion each year from its brand sales. That number is shocking enough on its own. But there's more to this story. Back in 2008, right after Mylan acquired the patent rights[2] to the EpiPen, allergy patients were only paying $103 for the *exact same prescription*. In the absence of any real market competitors or trustworthy generics, Mylan Pharmaceuticals had been able to increase the price for its most popular branded drug more than 500 percent in the span of just six years. Every few years, Mylan would raise the price by a couple hundred dollars, and no one had complained until that final price hike in 2016.

That summer, allergy patients and the parents of children with severe allergies took to social media to protest. What, exactly, were severe allergy patients without health insurance, or with poor coverage, supposed to do? The new price was prohibitive for almost everyone in the American working middle class and impoverished families. Since many schools are required to stock EpiPens for allergic students as part of the federally mandated School Access to Emergency Epinephrine Act, signed into law by President Obama in 2013, Mylan's pricing also put a strain on already strapped educational budgets around the country. People started purchasing vials of epinephrine and loading empty syringes with the same dosage, asking their healthcare providers to show them how to inject themselves or their children—a much cheaper, but riskier, alternative to EpiPen or other auto-injectors.

Despite the national outcry, Mylan kept the price of their EpiPens steady. Then, in the summer of 2018, allergy patients had to contend with a supply shortage. Many news stories reported panicked parents (the ones who could still afford to fill prescriptions) trying to find EpiPens in stock. Throughout the shortage, an EpiPen prescription remained priced at just over $600 for a standard two-pack—when you could find them at all.[3]

The story of EpiPen is an all-too-common morality play, highlighting the constant struggle between patient advocacy groups, doctors and other medical associations, and pharmaceutical companies. But if you think the story ends here, you'd be wrong. This tale gets even messier and murkier.

In August 2017, Mylan paid $465 million[4] in a settlement, after the Justice Department filed a claim against it for overcharging the federal government for EpiPen under its Medicaid program. That same year, Sanofi SA, a competing pharmaceutical company, filed an anticompetition complaint[5] against Mylan. The U.S. Securities and Exchange Commission (SEC) began investigating Mylan for price gouging. In 2019, Mylan agreed to pay an additional $30 million in a settlement with the SEC.[6]

While all this was going on, to help with access to lifesaving emergency adrenaline, the FDA approved the first generic version of the EpiPen and EpiPen Jr in August 2018. Manufactured by Teva Pharmaceuticals, the generic auto-injector pens work in much the same way as EpiPen; they have the same dosage and effectiveness and would be available for purchase in the following year. It's important to note that Teva's generic was not the first alternative adrenaline auto-injector on the market: Adrenaclick and Sanofi's Auvi-Q are two others that have been available for years. Auvi-Q is expensive without good insurance, coming in at a cool $598 for a two-pack (far less expensive than its peak of $5000 when I first began writing this book years ago). While Adrenaclick is much cheaper, at $109 for a two-pack, the injection mechanism differs from that of the EpiPen. That scares a lot of parents and patients, and even some doctors, since most people with experience giving adrenaline shots are trained on EpiPens. As one mother explained, "If it's an emergency, do you really want someone trying to figure out how to administer the cheaper version to your kid?" The Teva generic version was the first to duplicate EpiPen's injector mechanism, so it was likely to fare much better against the original.

Or at least it would have—if Mylan hadn't released its own generic first.

In anticipation of competition from Teva, Mylan began selling an unbranded version of its own medication at nearly half the price—$320 for a two-pack. Without at least a third viable generic on the market, Teva had to match that price (a two-pack is currently about $300) in order to stay competitive with Mylan's new generic. In 2019, the pharmaceutical company Upjohn (the generic arm of Pfizer) and Mylan merged to form Viatris, but the current pricing (as of this writing) remains the same—$650 to $700 for the brand-name drug and about $350 for the generic. The long marketing shadow of EpiPen as the brand to trust will also take a while to dissipate, despite the fallout of the pricing scandal.

What this brief history of the EpiPen scandal really exposes is the hard underbelly of a market-driven healthcare system. Unlike acute illnesses that might require one or two prescriptions or treatments over a short amount of time, chronic illnesses like allergy or diabetes require long-term maintenance and care. That means multiple prescriptions and treatments over the span of months, years, or even a lifetime. Food allergies are often for life (though not always), but auto-injectors expire and need to be replaced each year. They also cannot be stored at temperatures below fifty-nine degrees or above eighty-six degrees Fahrenheit, or the epinephrine inside will degrade, necessitating a replacement. That adds up to a nice, steady income stream for the corporations that manufacture them. Epinephrine is a cheap drug to produce (less than one U.S. dollar for 1 milliliter). The injector costs more to manufacture (estimates run from between two and four dollars, but the exact costs are unknown), and any small "improvements" on the drug delivery mechanism itself can justify a price hike, allowing companies to claim they need to recoup research and development costs. Obviously, drugs cost money to discover, test, and manufacture, but the larger social questions we should be asking ourselves are: How much should an allergy patient pay for a lifesaving drug? Who should cover that cost? The answers have serious consequences.

For example, Dr. Meenu Singh at the Postgraduate Institute of Medical Education and Research in Chandigarh told me that they don't have EpiPen in India. It's simply too expensive. Patients with severe allergies carry vials of epinephrine with them and try to get to a doctor when they have a food allergy attack. As a result, the death rate in India from anaphylaxis is much higher than it is in the United States (estimates range from 1 to 3 percent in India compared to 0.3 percent in the United States). The state of Illinois[7] was the first to require health insurance companies to cover the full cost of EpiPen for children who need it. In other states, however, parents often have a co-pay, if they have insurance at all. So, depending on which state you reside in, your access to an auto-injector differs. That means your risk of death differs. It's a simple and terrible correlation.

TALE #2: THE PROMISE—AND PRICE—OF DUPIXENT

In the five years it took to research this book, no allergy treatment came up in conversations more than Dupixent, the injection used to treat asthma and eczema. It dominated everyone's answer when I posed the question: What new treatments are coming online now that excite you? One of the answers was always, without fail, "Some of the new biologic drugs like dupilumab show great promise." For some, especially dermatologists, dupilumab's clinical trials gave them hope for a much-needed update to their typical arsenal of drugs. Its effectiveness at clearing the skin, they told me, seemed almost too good to be true. They were particularly excited to be able to offer it to their patients with the worst, most persistent, cases of eczema or atopic dermatitis.

Dupilumab—sold under the brand name Dupixent—is the product of a partnership between Regeneron Pharmaceuticals and Sanofi. The active molecule itself was discovered by Regeneron, with significant funding from Sanofi. In the late fall of 2021, after months of emails back and forth with both companies, I hopped on a video call with Dr. Jennifer Hamilton, vice president and head of Precision Medicine at Regeneron, and Dr. Naimish Patel, head of Global Development, Immunology and Inflammation at Sanofi, to talk about the development of their groundbreaking drug.[8]

As one might imagine, big pharmaceutical companies are often wary of talking to outsiders, particularly journalists or authors, and especially in the wake of scandals like Mylan's or the more recent OxyContin lawsuits. I can't say that I blame them; pharmaceutical companies are often simplistically painted as greedy enemies of affordable healthcare, driving up drug prices. The story I'm about to tell here, however, is more nuanced than that. It's a story about how scientists working inside pharmaceutical companies both rely on and then contribute back to the body of biomedical research that often initially takes place in not-for-profit laboratories.

Basic immunological research conducted on allergic pathways is performed slowly and meticulously over decades in various different academic and governmental laboratories scattered across the globe and heavily funded by national governments, including at the National Institutes of Health in the United States. In other words, public funding of research on the basic mechanisms of allergic response produces actionable information. That information is then made publicly available in scientific journals or on websites, both on principle and on purpose, since one of the overall missions and goals of the NIH is to advance human health everywhere. This is how basic science—especially biomedical science—is supposed to work.

However, that publicly funded research is then built upon by for-profit companies like Regeneron and turned into intellectual property. So federal money is used to seed advances in our scientific understanding of a subject like immunology. Corporate money is then used to turn those advances into effective and marketable treatments and, of course, large profits. The public benefits in turn—or so the reigning argument goes—by obtaining access to better medical treatments.

If there's a bad guy here, then it's this pernicious funding cycle itself. If and when the NIH's budgets are cut or held steady, as they frequently are, less money becomes available for further research in basic science. What we are left with is more applied science—or scientific research done solely to produce products, typically ones that are priced according to the amount of innovation that went into them.

It's not as if the research and clinical trials that corporations like

Regeneron and Sanofi conduct do not add value; they absolutely do, as we'll soon see, and they should be financially compensated for those contributions. But we need to be asking ourselves some critical questions: Just how much should pharmaceutical companies profit for turning basic science into a successful therapeutic treatment like Dupixent? And how does the quest for a profitable drug shift the ultimate goals of treatment?

This story begins in the late 2000s. The first step in drug development, as Dr. Robert Schleimer at Northwestern University explained in chapter 7, is to find an unmet need and then try to fill it. Researchers at Regeneron realized, after talking with academic and clinical peers at places like Northwestern and Mount Sinai, that there was a dearth of effective treatments for atopic dermatitis, also known as eczema. They also knew that the allergic skin condition severely impacted a patient's quality of life, and a new treatment with fewer side effects than their current options (steroid creams and immunosuppressants) might greatly improve their overall well-being. Eczema seemed like the perfect disease on which to start testing a new molecule they had discovered, which appeared to be capable of interrupting a key allergic pathway.

But to really understand the difference between using a steroid versus a monoclonal antibody like dupilumab to treat eczema, we need to take a brief scientific detour.

From a purely technical standpoint, dupilumab is a monoclonal antibody (an IgG antibody subclass) that targets interleukin-4 (IL-4), a cytokine produced by T cells and mast cells. (Cytokines are proteins produced by immune cells that affect other cells; in other words, cytokines are signals to other cells to turn on or off certain functions.) IL-4 is part of a key signaling pathway in type 2 allergic immune responses. Dupilumab prevents type 2 inflammation associated with both the IL-4 and the IL-13 allergic pathways by binding to the IL-4 receptor alpha on immune cells. In layman's terms, that means that the drug works to prevent cell signaling that can lead to a full-blown allergic response. The IL-4 pathway is involved in the expression of multiple atopic conditions—not just eczema—which is why dupilu-mab also works well on other allergic diseases. (We'll come back to this important fact in just a bit.)

Drugs that are commonly used to treat allergy, like steroids and immunosuppressants, are not as specific as monoclonal antibodies. These drugs are made up of small molecules that bind to multiple targets on a cell and might impact several pathways at once. The broader the impact that a drug has on the body, the greater the poten-tial for serious side effects.

Patel explained that steroids might do well at suppressing the in-flammation that troubles patients with eczema, but at the same time they also suppress the types of inflammation that help our bodies to fight off bacteria, viruses, or fungi. The same biological pathways that steroids target are also critical for things like bone growth and muscle maintenance. That's why steroids cannot be used for long periods of time: Long-term usage can lead to brittle bones, bone fractures, or an increase in the incidence of skin infections. It's easy to see why doctors and patients have a love-hate relationship with topical and oral ste-roids. They both help and hurt. The newer antibody treatments like dupilumab, Patel said, are exciting precisely because they often cause far less collateral damage.

"We want something that's specific," Hamilton said, "because we don't want side effects. The goal was not to be an immunosuppres-sant. The goal was to target, specifically, the part of the immune sys-tem that was driving allergic disease."

Dupilumab had already shown promise in the lab in mouse mod-els, but the true test was clinical trials in humans.

As soon as they began seeing the initial results, Hamilton knew that they had something very special on their hands.

"It was the itch data that blew us away," Hamilton recalled.

The reduction was dramatic and took place much faster than the research team had initially predicted. They thought that the drug would take weeks to months to reduce the size of the skin lesions and make a noticeable difference in the level of itch that patients were experiencing. They were wrong. Most patients experienced relief after only a week. And because of the narrow targeting of the antibody, there didn't seem to be as many side effects as other treatments for eczema would cause. For instance, there were no increased incidents of skin infections during the trials. The new drug even seemed to alter the skin's microbiome in a positive way.

"After dupilumab treatment," Hamilton said, "the level of *Staph aureus* [a type of bacteria that can live on the skin] is lower. And you have a more diverse microbiome on the skin, which is more normal."

The treatment works like this: Dupixent is delivered via one injection with a pre-filled needle every other week. In most patients, the treatment begins working at around twelve to sixteen weeks after the initial injection and needs to be continued indefinitely to remain effective. It is therefore considered a "long-term" treatment, meaning that its usage is prolonged and sustained.

In March 2017, based on the data from subsequent clinical trials, the FDA approved Dupixent for use in eczema patients. Two years later, the FDA approved Dupixent for use in moderate to severe atopic eczema in adolescents, significantly broadening its label. In May 2020, the drug label was changed yet again to allow its use in children ages six to eleven. In June 2022, the FDA approved the use of dupilumab for children ages six months to five years with moderate-to-severe atopic dermatitis whose disease is not adequately controlled with topical prescription therapies or when those therapies are not advisable. These approvals opened up a broad patient population for potential use.

The market for pharmaceuticals to treat atopic dermatitis/eczema was huge even before Dupixent was approved for public use. In 2017, the overall global atopic dermatitis drug market was valued at $6 billion per year. The largest share of that went to corticosteroids, but biologic drugs like Dupixent have the potential to topple them as the biggest drivers of revenue in the dermatitis market. A mere four years after its initial FDA approval, Dupixent was already a $4 billion per year drug (the total global market is around $6 billion per year). As dupilumab's drug labeling changes to cover more age groups and more conditions, that share will likely swell; by 2026, the global market for atopic dermatitis is expected to grow to more than $13 billion per year. That spells a lot of profit for the pharmaceutical companies trying to corner the market. Regeneron and Sanofi stand to rake in unbelievably large profits, especially since their drug patent will last for at least another fifteen years, longer if—or more likely, when— they tweak the drug itself or its delivery mechanism. Already, in June 2020, the FDA approved a pre-filled pen designed to self-administer Dupixent. While the drug itself is the same, the new delivery mechanism will reset the patent for the self-administered versions.

But Dupixent's potential customer base is even larger than all this. Clinical studies in asthma patients showed a substantial improvement in lung function and significantly lower rates of severe asthma than in placebo groups. Based on initial data, the FDA approved the new biologic for use in patients with moderate to severe asthma in March 2018. And in June 2019, it was further approved for use in chronic rhinosinusitis with nasal polyps (a chronic inflammation of the sinuses that causes growths in the nasal cavities—in other words, a very unpleasant condition). Two years later in October 2021, the FDA approved the use of dupilumab as an add-on maintenance treatment of patients ages six to eleven years with moderate-to-severe asthma characterized by an eosinophilic phenotype or with oral corticosteroid-dependent asthma. In May 2022, the FDA approved the use of dupilumab for patients ages twelve years and older with EoE, weighing at least 40 kilograms. As of this writing, dupilumab is in phase 3 trials for prurigo nodularis, pediatric eosinophilic esophagitis, hand and foot atopic dermatitis, chronic inducible urticaria-cold, chronic obstructive pulmonary disease with evidence of type 2 inflammation, chronic spontaneous urticaria (or hives), chronic pruritus of unknown origin, chronic rhinosinusitis without nasal polyposis, allergic fungal rhinosinusitis, allergic bronchopulmonary aspergillosis, and bullous pemphigoid. I do not have to tell you how much all these potential changes to the drug's label would expand future revenues for Regeneron and Sanofi—all from a single "miracle drug." From these clinical trials alone, I don't think it's a stretch to argue that dupilumab is being viewed by the medical community as a possible one-stop treatment for almost all allergy-related disorders. Patel told me that Sanofi is partnering with academic institutions to do longitudinal studies to see if early treatment with Dupixent can halt the atopic march, or the progression in children with eczema to developing asthma or food allergy. (From what I've seen, I wouldn't bet against that outcome.)

So far, all of this seems positive. Maybe, in reading this, you have also begun to get a little excited about the drug's effectiveness and potential promise for treating other allergy-related disorders. That is the road I initially traveled down while doing research for this book. And yet, over and over again, I kept hearing from the allergy experts I interviewed about the dangers of seeing any *one* treatment as "the cure" for any allergic disease. I was reminded again and again, from people with decades of experience working with allergy patients, that Dupixent, like every other pharmaceutical intervention, has its limits.

First, there are its side effects.[9] As we saw in the last chapter, allergists and immunologists are already concerned about the possible negative side effects of taking dupilumab, even as they remain excited about the possibilities of the drug's overall effectiveness. And the data, so far, backs up their concerns. In a study of a cohort of 241 French patients[10] using dupilumab, researchers found that while the drug was as effective as reported in the initial clinical trials, it also produced a higher rate of conjunctivitis (38 percent of patients) and eosinophilia (24 percent higher than at baseline before taking the drug). Eosinophilia is the term for having higher counts of eosinophils, a type of white blood cell. Having a higher number of eosinophils is typically associated with parasitic infections, some cancers, and allergic reactions. Another study[11] found that 23 percent of patients taking dupilumab developed "new regional dermatoses," or new patches of irritated skin, typically on their faces. The researchers suspect that these new skin flare-ups might be caused by an unknown underlying contact dermatitis allergy and recommended that patch testing be used before a patient goes on the new biologic. However, they also noted that not all cases could be explained by allergic triggers.

As an important aside, many patients don't seem to mind the tradeoffs, even when they are as severe as ulceration of the eyes. Patients in online forums and those who took part in qualitative research on Dupixent collected by the National Eczema Association argued that they do not want to stop taking the drug as long as it is effective at clearing their skin. Many sing the praises of Dupixent and urge others to get on the drug as soon as their doctors and insurance will let them. This is a testament, I think, to how much suffering moderate or severe eczema causes. Most people with a more severe form of the condition seem willing to trade any side effects for, as many put it, getting their normal lives back. They consider it a "life-changing" treatment.

And to be fair, that's why researchers like Jennifer Hamilton and Naimish Patel became biological researchers in the first place. They know that conditions like eczema can be terrible burdens, and they want to help alleviate suffering. Hamilton keeps an email from one of the initial trial patients for dupilumab on a bulletin board in her office, to remind her why she does her job.

"At the end of the day," she said to me, "we're all in this business to improve patients' lives and save lives." And, for many people, Dupixent does just that.

But, as I was reminded again and again by clinicians, there will always be patients with moderate to severe allergic disease for whom this "miracle drug" will not be effective. Clinicians report that about one-quarter of patients do not respond as well to Dupixent as initially hoped. According to the independent Institute for Clinical and Economic Review (ICER), 30–44 percent of patients show dramatic improvement on the drug, enough to justify its high cost. Yet there are others who seem to respond very well to Dupixent initially, but then revert to their original condition. In these patients, the drug's effectiveness wanes. As with other treatments, like corticosteroids, the body seems to adapt to them, making them less efficacious over time. These patients are typically labeled "non-durable responders."

If, as Dr. Kari Nadeau, the director of the Sean N. Parker Center at Stanford, worries, we become too enamored by the success of one drug like dupilumab, we might stop vigorously looking for other, better treatment options altogether. Or, worse still, we might stop trying to discover a permanent solution to allergic responses. That doesn't seem to be happening yet, as many new drugs—many of them biologics like dupilumab—are currently being evaluated for eczema alone. One of those drugs, lebrikizumab, is showing even better results than Dupixent in clinical phase 2 trials.

Finally, not everyone will be prescribed dupilumab for their conditions, which may be deemed "not serious enough" to justify the high cost of the drug. There is no agreed-upon clinical definition of mild to moderate or moderate to severe atopic dermatitis. As we saw with the definition and classifications of asthma, this is a common problem in allergy care and makes diagnosis more challenging. Without standard global guidelines and definitions, each practitioner is forced to use a combination of their own clinical experience and diagnostic criteria to determine the level of severity of a patient's disease. And diagnosis of a moderate to severe case, as Dr. Jessie Felton, a pediatric dermatologist in the United Kingdom, explained to me, is absolutely critical to getting access to a new drug like dupilumab. Because Dupixent is so new and so expensive, and no cheaper generic versions exist yet, most insurers and national healthcare agencies (like the United Kingdom's National Health Service) reserve coverage for only the most pernicious and severe cases. A moderate case might not be covered, especially if the patient responds reasonably well to corticosteroid creams or short bursts of immunosuppressants. Felton said that the hoops her patients have to jump through to get access to Dupixent are many and difficult. In other words, Dupixent might be amazingly effective in treating multiple allergic conditions, but only for those who will be able to access it. And for now, that number is still limited.

For years, I've lurked on various social media sites, observing the interactions of people with allergy. In atopic dermatitis and eczema discussion threads, there is palpable excitement about the promise of Dupixent. Some lucky patients even post dramatic before and after photos to demonstrate the drug's seemingly miraculous ability to clear up terrible flare-ups. But all too often, in the same thread, there are people complaining about their health insurance companies not covering the drug for their own care. "I'm so happy for you!" wrote one commenter. "I wish I could get Dupixent, but my insurance company won't fucking cover it because they said my eczema is 'not adversely affecting my life enough.'" Others share tips about how to take advantage of the programs that help people afford the co-pays and access the drug without insurance. (All of this information, by the way, is readily available on Sanofi's official site for Dupixent.)

The high cost of the drug—currently $3,203.39 for a four-week supply without insurance—is out of reach for many patients who might benefit from using it. For those of you who don't want to do that math, at those prices, a year's supply of Dupixent comes out to $41,644.07.[12] Approximately 80 percent of health insurance companies will cover Dupixent, but the co-pays range from $60 to $125 per month. For someone on a strict budget, the co-pay might still be unaffordable. And what about patients on Medicare or Medicaid? Only some of the plans will cover Dupixent at all. Patients in less wealthy countries will likely have little to no access to Dupixent for at least a decade. And yet, the growth in revenue streams for Dupixent is only limited to the number of patients who will eventually need the drug. And there are, as we've seen in prior chapters, going to be a lot of them.

Regeneron sees patient and physician education[13] about dupilumab's success rates as crucial to growing the drug's market share of dermatitis pharmaceuticals. From their perspective, since patients had so few options for so long, they need to learn that dupilumab is not only available, but much better than corticosteroids or immunosuppressants. If this sounds familiar to the EpiPen story, it is. At least in the United States, direct-to-consumer marketing of new drugs is a big part of the modus operandi of all pharmaceutical companies. If they can get the word out about a drug's effectiveness, then patients will ask for that drug by name in their physician's office. As a medical anthropologist who is familiar with patient-doctor interactions, I can tell you that most doctors—in addition to wanting to promote the basic health and wellness of their patients—feel a great deal of pressure to make their patients happy. In a capitalistic system, a patient is akin to a customer. As CEO of Regeneron, Len Schleifer needs to make his company's shareholders a profit as well as keep *his* customers—patients and insurance companies—satisfied. For his efforts, Regeneron's board of trustees agreed to $1.4 billion in bonus compensation for him if the numbers keep growing at the same pace. So far, Dupixent is the fastest-growing biologic drug ever to hit the market. And with more and more people developing allergic disease, it shows no signs of slowing down.

The story of the development of a drug like Dupixent does not have an easy takeaway. It highlights the ways in which pharmaceutical companies build upon basic science performed in nonprofit research laboratories around the world. That may seem like an unbalanced exchange, especially since companies like Regeneron and Sanofi stand to profit greatly from that information. But the truth is that academic institutions do not have the resources (personnel, global connections, financing) to do the large clinical trials required to safely bring a drug like Dupixent to market. In turn, large companies like Sanofi do not have the time and resources to invest in long-term studies to better understand things like disease progression and disease determinants. That's what academic research labs do well. Academic, government, and pharmaceutical scientific researchers are in a dynamic, complicated relationship. And as we've seen here, it works. But it also *doesn't* work, at least in terms of equal access to research funding and the cost of most new treatments. The tension is between wanting to help patients and needing to make a profit, as we'll see in the very next tale.

TALE #3: ACADEMICS IN BUSINESS

Some of the immunological scientists and clinical researchers I spoke to for this book have two jobs. On the one hand, they are researchers ensconced inside universities and/or teaching hospitals. Their official positions are as professors, clinical researchers, or laboratory scientists (and, in many cases, all three at once). On the other hand, they are novice businesspeople who have leading positions inside new biotechnology start-ups. This is not an unusual pairing within the modern university; academic institutions are increasingly partnering with corporations to fund research labs on campus. In many ways, this makes a lot of sense. The allergy experts I interviewed are all, without exception, at the top of their fields. They know, in aggregate, all there is currently to know about allergic-type immune responses. Any biotechnology company looking to churn out new or better diagnostics or therapeutics could find no one better suited for developing innovative products.

There is also a palpable tension in these arrangements, however, and one that the researchers I spoke with often talked about directly and openly. The inherent conflict between a biomedical expert's desire to help patients and a company's need to make a profit is present at all stages of research and development. The academics who choose to sign on tend to have one overarching goal: to create better things that help more people. That's what most of them hope to accomplish by becoming a part of these business ventures. And yet any explicit mention of the money at stake in these arrangements between academics and corporations was often avoided or elided altogether. Academic salaries are solidly middle-class; science and engineering faculty make anywhere between $90,000 and $150,000 per year, depending on their seniority and their institution (Harvard obviously pays more than the University of Nebraska). I imagine, for many academics and physicians who try to bridge the gap between basic and applied scientific research, or between the university and the business world, it can be exceedingly difficult to strike a balance.

Because of this, some experts admit that they try to stay clear of the profit motive altogether. Dr. Marc Rothenberg at Cincinnati Children's Hospital is wary of the hype machine that so often goes along with biotechnology companies, with their reliance on garnering venture capital investments rather than being motivated by the science and needs of patients. Doctors and basic researchers mainly worry about the health of their patients and how to improve their lives, through clinical care and research. Academic labs, such as Rothenberg's, are intellectual property engines, fueling the pharmaceutical company with good ideas.

"Things will happen," Rothenberg said. "The discovery of a new pathway with translational significance often triggers companies to get interested. Now one of the hottest topics of companies and venture capitalists is focusing on drugs that are going to block eosinophilic diseases like EoE. I get calls every week, if not every day. Ten years ago? There was barely any interest."

For Rothenberg, the patients' well-being comes first. If his own research can help someone, then he's done his job. But he remains wary of letting corporations decide which avenues of research to pursue, since those are typically the types of research they can monetize.

Dr. Dean Metcalfe at the NIH explained to me that the process is almost always the same. The inhalers and steroids and biologicals used in clinical settings necessarily come out of big pharmaceutical companies because they have the money and systems in place to do research trials. People in academic labs will work on something like signal transduction in allergic response using NIH funding, but then won't have the resources—staff and funding, typically—to develop new molecules (like dupilumab) that might affect that signaling. That's where the pharmaceutical companies step in. The academic research on signaling is published in scientific and medical journals and then the corporate labs take those findings and start looking for the molecules that will turn those signals on and off, as needed.

"The challenge today is that it's so damn expensive to do all that," Metcalfe suggested.

Just how expensive? It's hard to tell with any exactitude. For one, getting any detailed information about the costs associated with the research and development of new drugs is like getting gold out of Fort Knox. Pharmaceutical companies aren't exactly forthcoming about their costs, and estimates range widely: from $19 million per FDA-approved drug to nearly $3 billion. Either way, that's money that the National Institutes of Health can't afford to spend. Its total fiscal budget in 2022 was $46 billion (spread out over many different research centers and with all diseases and health conditions *combined*). In other words, academic researchers who rely on NIH funding often need to look elsewhere to make research ends meet—and that's why outside funding from corporations can be so enticing. Most nonprofit laboratories can't survive without outside funding of some kind. Almost every academic researcher has to navigate the choppy waters that divide basic science from its applied—and more profitable—counterpart.

One of the most reluctant participants in the business side of allergy that I interviewed was Dr. Cathryn Nagler at the University of Chicago. When we spoke at length about her research in her office on the university's campus, she explained that she had to be courted by entrepreneurs. Initially, she saw herself as just a scientific researcher, an immunologist on a mission to better understand the human microbiome. Of course, Nagler always wanted that research to have positive effects for food allergy patients, but that was a side effect of her attempts to better understand the interplay of microbes and immune cells in our guts. Then, slowly but surely, that side effect began to move to center stage.

"It's not just academic research anymore for me," Nagler said. "Now I want to do something to help the people that contributed to all this research. I want to fulfill my promise to them."

Working with a colleague in Naples, Italy, Nagler and her team took samples from healthy and cow's-milk allergic (CMA) infants, looking for key differences in their microbiomes that might explain their conditions. They transferred these microbiomes into germ-free mice, which have no bacteria of their own, then looked at the changes in gene expression induced in the intestinal epithelium of the healthy and CMA-colonized mice. When they integrated data sets on the differences in the healthy and allergic microbiomes and the genetic changes they induced in the mice, they found that one type of anaerobic bacteria in the Clostridia class, *Anaerostipes caccae*, was significantly more abundant in the healthy infants. Nagler thinks this particular class of bacteria is one of the "peacekeepers" of the gut. They ferment dietary fiber to make short-chain fatty acids like butyrate, which is essential for the health of the epithelium, or the thin protective layer of our guts. *A. caccae* also induces or creates regulatory T cells and regulates the gut barrier.

The first product of Nagler's company, ClostraBio, is a synthetic polymer that targets the delivery of butyrate to the parts of the intestine where it is normally produced. She is also interested in the potential development of live biotherapeutics and prebiotic dietary fibers to promote the growth and well-being of bacteria like *A. caccae* in our guts. Nagler's co-founder is a molecular engineer, and Nagler tells me that, through the process of working closely with him on the business venture, she is now converted to the idea of translational research.

"We're going out for series A funding, which is twenty million dollars, to take this into the clinic," Nagler said, showing me slides of the initial findings. "I want to translate my academic work into treatments."

Unlike many other treatments that need to be taken continuously, Nagler expects that her therapeutic will not be lifelong. She hopes that ClostraBio's therapy will help to restore effective barrier function to promote immune tolerance in food allergy patients. And the best part, for Nagler, is that her therapeutic will alter the gut microbiome for the better—without using antibiotics, which are a big part of what causes the problem in the first place.

For Nagler, the importance of crossing the divide between basic science and applied science was in making a difference for the people who have contributed to her research over many long decades: the patients themselves. Although she was—and remains—hesitant about turning science into a profitable treatment, she also sees how doing so can help a lot more people. And what else, really, is scientific research for if not to aid the health and well-being of all humankind?

MONEY, MONEY, MONEY, OR, HOW TO FUND BASIC SCIENCE

It's not just pharmaceutical companies raking in the dough from allergy sufferers. The air purifier market, for instance, is anticipated to reach $28.3 billion by 2027.[14] Hotel chains have begun to capitalize on the burgeoning allergy market, too, by providing costlier "allergen-free" rooms.[15] Allergy-friendly or hypoallergenic products are everywhere, despite little to no regulation on what can be labeled as "hypoallergenic."[16] The management consultant firm McKinsey & Company estimates that there are currently eighty-five million American shoppers who avoid at least one of the major food allergens and who are willing and used to paying more for "safe" foods. All of this adds up to big revenue for the corporations who cater to the growing number of allergy sufferers worldwide.

As we've seen, however, therapeutic and preventative treatments are where some of the biggest money is to be made in the allergy markets. Drugs like Dupixent can make billions of dollars for their manufacturers at the same time that they make a sizable portion of allergy sufferers feel better. The trick is in finding a balance between the need to advance treatments and find possible fixes for allergic responses, and the desire to make a big profit from doing so. The uncomfortable conclusion is that we need to be extremely careful about how we spend our research dollars and about who conducts clinical studies.

When I sit down with Dr. Alkis Togias, the branch chief of the NIH's division for allergy, his mind is very much on money and its role in promoting the type of scientific discoveries that ultimately lead to better treatments. It's a complicated issue, to be sure, and one that Togias is all too familiar with.

"The NIH is by far the biggest funder of allergy research in the country, and in the world," Togias says, laying out the problem of research funding versus profit. "We are advocating, we're funding specific research groups for food allergy, for asthma, but we're always bringing them together into the bigger picture. We're always supporting research into the basic mechanisms of allergy as well as clinical allergy research. It's our inherent belief as an institution that basic science is going to be extremely important in understanding disease."

Togias gives me an example of what the NIH can do that no one focused on making a profit will do. When the NIH funds a clinical trial, it insists that the study include an understanding of the mechanisms of the disease itself. It's not enough, in other words, to prove a drug works. The NIH wants to know exactly *why* that drug works.

"People will say, 'Well, why do you do that? You're testing a new drug; if it works, that's great.' Our answer is, well, it is great if it works, but it's by no means going to be the ultimate treatment," Togias explains. "We need to gather information that will tell us something about the next generation of treatment, about the next steps we need to take. No pharmaceutical company is ever going to do that."

Even donors and patient nonprofit organizations don't care why something works, just that it does. The emphasis is on the result, not the basic biological mechanisms. But the problem with that approach, as Togias points out, is that we won't expand our knowledge base. And then we won't have as good of a chance of understanding how our immune responses work and why, or be able to alter them when it really counts—*before* an allergic reaction ever takes place. Funding basic science research that attempts to better understand how our immune systems function, like current NIH research on how mast cells release histamine or the Immune Tolerance Network, is probably a much better use of our limited resources than trying to find another, even better, drug or product that merely prevents or treats our symptoms. Ultimately, discovering a method to prevent all allergic reactions from ever developing in the first place would be worth a hundred new biologic drugs or treatments. And only a reinvigorated governmental and social investment in basic immunological science—and with it a societal shift that starts to decouple monetary profit from healthcare—can get us there. Especially if "us" includes everyone, not just the wealthy, first-world, white, and/or urban allergy sufferers who can afford to pay for the best treatments.

CHAPTER 9

What Makes a Treatment Effective? Weighing Benefits and Risks

Something that we haven't really discussed yet, as we've explored allergy care, is how people—allergy patients, caretakers of children with allergies, practitioners and clinicians—think about and weigh different treatment options. At the heart of these decision-making processes are debates over perceived costs and benefits, a particular treatment's reported effectiveness and possible negative side effects, and the overall safety and well-being of the patient, both physically and mentally. Patients who undergo treatment for more severe immune reactions often need to accept a certain amount of risk. The solution for one ailment almost always has the potential to cause another, especially when we're talking about something as finely balanced and inscrutable as the human immune system.

I'd like to pause here and ask you to engage in a brief thought experiment. You may or may not have moderate to severe allergic conditions yourself, or you may or may not know someone who does. You will have varying degrees of familiarity with the issues we're going to discuss in this chapter. So to start, I'd like to get us on the same page about what treatment decisions can look like, and what can be at stake.

I want you to imagine yourself as the parent of a five-year-old with a serious food allergy to peanuts. Your son's allergy is so severe that he might die simply from a trace exposure. Every time you take him to a birthday party or a playdate or drop him off at school, you worry. You have become a broken record, repeating the warnings about his allergy to everyone who comes into contact with him or has to care for him in any way. The constant need for vigilance and the low levels of everyday anxiety are exhausting. Your five-year-old has started to develop anxiety, too, as he has become more aware that invisible things in his environment can harm him. In fact, your entire family life has been impacted by your child's health condition, since keeping him away from peanuts is almost a full-time job in and of itself. This has been your daily reality since you first discovered his allergy more than four years ago.

Now I want you to answer a question based on this scenario: If all it would take for your son to have a deadly anaphylactic response was a family trip to the local ice cream stand, where a new teenage employee unthinkingly used the same scoop they used to serve someone else Reese's Peanut Butter Cup on your son's bubblegum ice cream, would you elect to try oral immunotherapy (OIT) to treat his food allergy? Even if it might cause the exact same reaction?

That's an extremely tough question to answer, any way you look at it. But what if your allergist and your pediatrician were split on their recommendations, one insisting that the new OIT treatment shows great results, and the other telling you he wasn't comfortable recommending it for your child because not enough studies have been done yet on its long-term effectiveness or on the rate of adverse events during treatment? Then imagine that you spend hours at the computer doing your own research, fixating on the small percentage of kids who go into anaphylactic shock during the treatment itself. Is the risk of a few accidental exposures over his lifetime preferable to the risk of knowingly—and for months—exposing your son to a minute amount of allergen that triggers anaphylaxis?

If he undergoes the treatment and it is successful, then your son can accidentally consume the equivalent of two nuts without ending up in the emergency room. And that sounds incredible to you. But then you also learn that since OIT is still very new, we don't yet know if the positive effects of the treatment will last more than ten years, and that maintenance therapy has to be continued indefinitely. If he stops the treatment, the small amount of immune tolerance that he gained through a grueling and stressful series of treatments might be slowly lost anyway. And this all assumes, of course, that you have good insurance and access to an allergist and can afford the cost of the co-pays for each office visit.

Now answer the same question again: Would you risk your child's life to prevent him from ever experiencing an anaphylactic event, or not?

These are not purely hypothetical questions. They are very real, and they're being asked by parents of children with severe allergy every day. In talking to people while researching this book, it became clear to me that clinicians, patients, and parents weighed the possible risks and costs of treatment very differently. It was not a "no-brainer" decision to undergo OIT, for instance—far from it.

There are big moral and existential questions at stake in these decisions, including how much risk a person living with a severe or life-threatening allergy might be willing to endure in order to achieve a "cure" for their condition, or at least alleviate their worst symptoms. FDA regulations ensure that no biological therapy can cause undue harm to its patients, using a Benefit-Risk Framework. According to the FDA, the framework "is a structured, qualitative approach focused on identifying and clearly communicating key issues, evidence, and uncertainties in FDA's benefit-risk assessment and how those considerations inform regulatory decisions."[1] In other words, regulators use data from clinical trials to weigh the benefits and risks to patients. Yet even after a new treatment gets FDA approval, like Aimmune's new drug Palforzia for peanut allergy, it doesn't necessarily mean that allergy experts or patients will agree that the benefits outweigh all the risks.

The truth is that treatments for any allergy don't always work for everyone and their effects—even when the treatment *is* effective—don't always last. Treatments can also be costly and difficult to maintain because they need to be continued for years (sometimes for life).

In what follows, we'll see how different stakeholders define what an "effective" treatment is in relationship to food allergy and atopic dermatitis, and how patients navigate their treatment options in relationship to the development of new FDA-approved therapies. It's a messy story, as always. At the end, the choice to undergo a new treatment is personal and based on someone's lived experiences. In the absence of any real preventative therapies, many patients with the worst allergic conditions are left with only two real choices: to experiment with new treatments . . . or not.

REAL-WORLD EXAMPLE #1: ORAL IMMUNOLOGY
TREATMENTS FOR FOOD ALLERGY

Before we begin, some background: As we've already learned, immunotherapy itself is an incredibly old idea. Allergists have been practicing it for well over one hundred years, to middling effect. The underlying principles of the newer, more standardized, and advanced immunotherapy treatments coming to market now, like Palforzia for peanut allergy, remain the same. The ultimate goal of immunotherapies is to retrain the immune system to better tolerate allergens. Currently, we have immunotherapy treatments for some respiratory and food allergies, and they vary in their effectiveness depending on the individual being treated. It's not an exact science yet; from a strictly biological standpoint, we're not even certain how immunotherapies work. We just know that they do in many cases.

In the past, immunotherapy for food allergy was mostly a DIY affair. From the early 1900s to the 1970s, allergists typically made their own allergen extracts using local pollen or other locally sourced allergens. Throughout the 1970s and '80s, as the science advanced, allergens were standardized and manufactured to different specifications. Today, allergens are available to order, and allergists mix and dilute[2] them to suit an individual patient's needs. Each allergist will decide the concentration of allergen for each patient and protocols can vary widely.

For peanut allergies, specifically, both early and present-day allergists would often buy peanut flour in bulk, using it to produce their own sublingual or oral immunotherapy tablets that gradually increased trace amounts of the allergen over weeks and months. These homemade peanut pills were, as several allergists told me, incredibly cheap to make. The process was also relatively simple, despite requiring specialized knowledge to ensure the correct amount of allergen went into each dose. Since the pills were not standardized, however, mistakes in dosing were possible and sometimes occurred. Today, in addition to this method (which is still available), patients can also choose to take Palforzia.

Palforzia is a prescription medicine that is made of defatted peanut flour. It is currently indicated as a treatment for patients ages four and up, though the treatment works better in younger children than it does in adults. The treatment is taken daily for six months and is completed in three phases, from the initial dose escalation, up-dosing, and maintenance. The initial dose contains 3 milligrams of peanut and gradually increases until the final dose of 300 milligrams. At this point in time, Palforzia is the only FDA-approved oral immunotherapy prescription; it is only used to treat peanut allergy. Patients initially undergo supervised treatment in a clinic in case they experience a significant adverse reaction. Each up-dose (which contains a slightly larger amount of the allergen) is administered by a professional under supervision, but if the patient responds well, other doses are taken at home.

Oral immunotherapy treatments for food allergy—both old and new—have risks. In a patient with severe allergy, the treatment itself can cause them to experience an anaphylactic event (which is why all initial doses and up-doses must be administered in a clinic or hospital with access to lifesaving equipment and expertise). Even when the patient tolerates the treatment, it can—and often does—cause discomfort. Side effects of OIT (both Palforzia and traditional OIT) can include tingling or swelling of the mouth and tongue; difficulty breathing or wheezing; throat tightness or swelling; swelling of the face or eyes; skin rash or itching; stomach cramps, vomiting, or diarrhea; and dizziness or fainting, though the dosage can be temporarily scaled down to alleviate the worst symptoms. In some patients, the treatment can cause inflammation of the esophagus or trigger EoE. Many patients (and their caretakers) also suffer from heightened anxiety about these possible side effects during treatment, especially in its initial phases or during up-dosing. Some will quit treatment due to one or more negative side effects.

In a recent study of 1,182 individuals (ages four to seventeen) taking Palforzia, most experienced mild (35 percent) to moderate (55 percent) symptoms during the initial weeks of treatment.[3] Serious reactions occurred in forty-one patients, or 3.5 percent of people taking the drug. Anaphylactic events were rare but did occur at the rate of 1.2 percent over a three-year period. They usually decreased in frequency as treatment progressed. The main side effects reported were throat irritation, stomach pain, and itching of the mouth. Three out of four patients taking Palforzia were able to achieve the final maintenance dose of 300 milligrams.

The decision whether or not to undergo some form of OIT often seems to rely on different ways of thinking about the treatment's overall effectiveness. The definition of "effective" in relationship to different stakeholders can vary dramatically.

Viewpoint #1: The Patient's Perspective

Stacey Sturner runs one of the largest food allergy groups on Facebook. She began it in 2015 when her family was dealing with her son Reid's peanut allergy, which he had developed at twelve months old in 2013. The Facebook group provides information and a shared space for empathy, support, and personal stories. Most people, Stacey tells me, join the group because they're in the midst of a treatment decision—like whether or not to go on Palforzia. Other people join once they've begun oral immunotherapy, sometimes to brag or to shame other parents for not making the same decisions, and sometimes to share the negative experiences they're having on the treatments and looking for support to navigate them. As a communications and marketing professional, Stacey wanted her Facebook group to go beyond what she calls the anecdotal approach.

"Most social media groups approach the problem of allergy anecdotally. 'What is your story and how does your story apply to me?' And I think that is problematic for various reasons," Stacey explains during our second long phone conversation about allergy. Our discussion is specifically focused on how people like her make treatment choices.

Stacey tells me that she was lucky enough to have a fantastic allergist when her son was first diagnosed—and that others are not so fortunate. She thinks that up-to-date, evidence-based information is key to understanding allergy and its treatments. Good information can make all the difference between living well with allergies, or not. When she first started participating in online groups, Stacey quickly grew frustrated at the lack of evidence-based information being shared. There were a lot of anecdotes and "facts" that weren't backed up by any citations or links. By January 2014, Stacey was doing so much research herself that she decided to start her own group to share it. Her group now has more than thirteen thousand members, but they are highly curated: Stacey doesn't let anyone stay in the group who refuses to back up their posts with scientific data.

"There are a lot of desperate families out there," she says. "Food allergy can have a detrimental effect on your lifestyle. There are a lot of mental health issues because of all the stress."

In 2016, Reid failed an oral food challenge for peanuts in the allergist's office. The reaction was, thankfully, relatively mild, but still required a shot of epinephrine. In fact, Reid has never had an anaphylactic response and has never been to the emergency room from an accidental exposure. His IgE antibody levels in response to peanuts have always been on the lower end of the spectrum, and he was expected to naturally outgrow his allergy. Because of this, Reid wouldn't have qualified for any of the clinical studies for Palforzia that were ongoing at the time. Her son's condition was moderate enough that it would have been fine for Stacey to take a "wait and see" attitude. Still, for her his reaction was big enough that she didn't want to take any chances. She gave herself a deadline: If by age five Reid still reacted to peanuts during a food challenge, then she would try OIT before he began attending kindergarten.

Stacey tells me that Reid's relatively moderate reactions during food challenges ultimately made her decision easier in two ways. First, she could visibly see that he had an allergy (he would get a skin rash and swelling) and that he could benefit from the treatment. Second, she had a lot of confidence that he would be able to cope with the treatment itself, which required him to consume small amounts of peanuts for weeks on end. Reid, unlike many children who have learned to fear food as dangerous or deadly, didn't have any food phobia. Stacey had always been careful to explain his condition to him in measured terms, precisely because she had never wanted to terrify him unnecessarily or make him afraid to eat. As a result, Reid was more willing than many allergic children to consume something that he knew contained peanuts. Because of their unique situation, and armed as she was with the latest information on success rates and potential risks, it was much easier for Stacey to make the decision for Reid to undergo OIT.

There was, however, one thing that gave her pause. Her oldest son has Crohn's disease, an immune disorder that affects his esophagus. (Crohn's typically only affects the bowel, but in rare cases can also affect the esophagus.) Oral immunotherapy is not recommended in patients who have eosinophilic esophagitis (EoE), since it can trigger their disease and cause serious—often life-threatening—complications. The last thing Stacey wanted was to cause Reid to have additional issues caused by OIT.

"I think anyone would say that the treatment should never be worse than the disease," Stacey says.

Ultimately, her oldest son's specialist, an expert on EoE, okayed her decision to go ahead with OIT treatment for Reid, despite being "lukewarm" to the idea. Stacey went back to the specialist who had performed Reid's oral food challenge and signed up for OIT. During the initial treatments, when she and Reid would go into the clinic and take the dose and wait for forty-five minutes, Stacey would be a little nervous. But over time, as she saw Reid tolerating the treatment with little negative response, she felt more confident. She tells me that many parents in her Facebook group report similar experiences; as their children start making progress, their anxiety reduces, and their excitement and hope increase. That's when allergy experts need to inject a dose of reality, Stacey argues, to keep the parents' expectations in check. Just because a patient responds well to OIT initially doesn't mean that the rest of the treatment will go as smoothly. There is, as we've seen, a lot of variation in an individual's immune responses, so OIT outcomes can also vary greatly.

In the end, Reid participated in standard OIT with great success. The biggest hurdle was that the treatment disrupted the normal flow of life, since his therapy needed to be taken twice a day for months. That schedule was, by Stacey's own account, hard on both her and her son. But by six months, Reid had built up a tolerance to eight peanuts, or approximately 4,000 milligrams[4] of peanut protein. After that, he went on maintenance therapy for three years, taking regular doses to keep up his tolerance level. Eventually, his antibody blood levels dropped to near zero, and his skin-prick test was negative. At that point, his doctor suggested that Reid discontinue the maintenance therapy for a month and come back in for another oral food challenge. He passed it with flying colors, eating fourteen peanuts without a reaction. Now Reid eats peanuts at least twice a week to maintain his tolerance, which he is more than happy to do. Stacey tells me that Snickers candy bars are a particularly favored "medicine."

"We couldn't be more thrilled, and of course it makes me an immediate fan," Stacey says. "Except I have to be really careful, because I do know that the vast majority don't end up with such a great outcome. It's super tricky—I'm running a food allergy group, and I don't want people to use my story as an example of what could happen to them. Because, yes, it could. But the chances are not as high. We got extraordinarily lucky."

Stacey tells me that one of her concerns now is the lack of consistency in many OIT protocols. Each allergist does it differently, and that creates a lot of confusion and fear in food allergy parents. Because of her own research, she felt more confident about enrolling her son in OIT treatments. But she also understands why every parent can't or won't make the same decisions.

Since Palforzia's approval in January 2020, Stacey has seen an uptick in people joining her group who are undergoing OIT with the new drug. She sees this as a positive development overall, since Palforzia seems to be making OIT at least partially more accessible, even in areas without great access to allergists. Because Palforzia is a standardized protocol, with a documented success rate, more doctors and patients seem more comfortable with it. It has been, Stacey argues, great public relations for the idea of oral immunotherapy itself.

Because Stacey is not only an allergy mom but also a food allergy advocate, I ask her to weigh in on what she sees on social media, and what she hears from other allergy patients and their families. Since she has both been through the process herself and runs a Facebook group where people are actively crowdsourcing their treatment decisions, Stacey is in a unique position to talk about how people think about risk. How do average people navigate the decision-making process?

She doesn't hesitate with her answer. The biggest problem, as she sees it, is that many people come to her group without any information about OIT at all. They've been told that their child is a good candidate, and they're thinking about undergoing treatment, but they have absolutely no idea what the risks are.

"I'm constantly amazed at how many people post comments along the lines of, 'Hey, we're starting OIT tomorrow. What are the risks?' I just can't get over that," she says.

In general, Stacey observes that people haven't had their expectations tempered by the evidence. They don't know, for instance, that OIT can produce gastrointestinal complications. They haven't been told that children with severe asthma shouldn't undergo treatment. The worst part, Stacey explains, is when they get upset after learning about the possible risks from more informed group members.

She thinks that allergists need to do a better job of explaining all the risks and benefits to each patient. Going further, all family members—not just the parent or caretaker most responsible for medical decisions—need to be on the same page about treatment. She tells me that if her son had experienced stomach issues, her husband would likely have pulled him out of the treatment. He just wasn't comfortable with making his son uncomfortable, and Stacey let her allergist know that up front. But all too often, that doesn't happen.

"Overall, there is a lot of misinformation about what oral immunotherapy is," Stacey says. "And so much of it is marketing jargon. I'm a marketer, so it drives me a little bit crazy. I just don't think the medical field should be dabbling in marketing ploys to get patients to choose a certain treatment that, of course, the company that makes the treatment financially benefits from. I do think there's a lot of good that comes out of OIT. But I also think patients deserve a better education going into it."

When everything is said and done, Stacey tells me that the decision to undergo OIT—via either Palforzia or more traditional treatments—is an individual one.

"I'm a fix-it person," Stacey says, explaining her own decision to treat her son with OIT. "But it's a personal choice," she said, and everyone has different levels of comfort with the possible risks of the treatment.

Viewpoint #2: The Expert's Perspective

Dr. Cathy Nagler at the University of Chicago understands patients' concerns about oral immunotherapy and the difficulty of their decision to undergo a treatment like Palforzia. If a patient goes through all of the physical, emotional, and time commitment required for OIT to work, but then has to worry about its long-term effectiveness, is that a good bargain? Is it worth it?

Nagler is not so sure. Or, at least, she doesn't think that Palforzia or other oral immunotherapy treatments should be the only thing clinicians offer their patients. That's because oral immunotherapy isn't a "solution" to the real problem—which is the underlying allergy.

"What I'm going to argue to you is that it's never going to be good enough," Nagler said. "Oral immunotherapy aims to turn off the immune response. But we want to improve the bacteria-induced barrier response to keep the allergens from having so much access to the bloodstream in the first place. I think that if you don't do those side by side, you're never going to get better than transient desensitization. Even if you maintain it for life, that still may not be good enough."

And there is nascent research to back this up. A study performed at Stanford University in 2019 found that any discontinuation of peanut OIT therapy, or even a continuation at a lower dose, led to a significant decline in tolerance.[5] In the study, participants who had passed an oral food challenge after twenty-four months of OIT therapy were given either a daily dose of 300 milligrams or a placebo. Then, after a year, all the participants were given another oral food challenge. The results? Around 37 percent of the group that had been given the maintenance treatment passed the challenge for peanut; only 13 percent of those in the placebo group did. What this indicates is that discontinuing OIT maintenance doses reduces the protective desensitization. It also suggests that even if a patient adheres strictly to the maintenance schedule, they will probably still react to the food. It likely won't kill them if they consume a small amount, but they still need to avoid the food.

Patients trying to decide whether or not to undergo OIT treatments are not necessarily always aware of these outcomes. They don't know that even if the treatment is successful, they still need to avoid the food. They also may still need to carry an EpiPen in case they have a bad reaction in the future. OIT is not a perfect long-term solution to food allergy, despite the hype and hope around it. While it may relieve families' biggest fears in the short term, there are still very real questions about its safety in the long term. In fact, in its report on Palforzia in 2019, the Institute for Clinical and Economic Review (ICER), a nonprofit organization that provides independent analysis of the clinical benefits and cost-effectiveness of medical treatments, noted "considerable uncertainty about the long-term outcomes"[6] for the drug as one of the reasons it unanimously rejected recommending the treatment.

Something that makes these calculations more difficult is the uncertainty surrounding how an individual patient may react.

As Nagler explained, reactions to allergens are completely unpredictable. They can range from mild (swollen lips, hives), moderate (stomach pain), to severe (cardiovascular shutdown and anaphylaxis). What's more, the reactions can change. On one day, you may have hives. On another day, you may have a more severe response. (Symptoms of allergy can vary due to the amount and type of allergen as well as the method of exposure.) Even measuring IgE antibody levels isn't always adequate to predict the severity of the reaction. A patient can have very low levels of IgE and yet still have an anaphylactic response at challenge. The opposite is equally true. A patient can have very high IgE and have no symptomatic disease at all. This doesn't make it easy for patients and parents to make choices about OIT and can cause high levels of anxiety, especially during the first phase of treatment, when reactions are more common.

For Dr. Scott Sicherer, the director of the Elliot and Roslyn Jaffe Food Allergy Institute at Mount Sinai, the debate over the "effectiveness" of OIT centers on the definition of a successful outcome—and that depends on how we understand desensitization in the first place. In the ICER final report, the panel of expert reviewers noted that "desensitization" as a concept is not well defined. Is a person "desensitized" when they can consume two peanuts or when they can consume thirty? Is it when they have low IgE levels or a negative skin-prick test? There is no consensus about what a term like "desensitization" actually means on a clinical basis. This can make it difficult to compare studies of OIT that use different standards for tolerance.

"Usually they require that you react to less than a third of a peanut to be in the study," Sicherer explained. "They might set a threshold of two peanuts. So if you can eat two peanuts at the end of the study, then you're a success. Let's say two-thirds of the people in the study who got the real treatment were able to do that at the end. So we have raised their threshold and that's what we are calling success. The group that was in the placebo group, they were the same as when they started. You could look at it that way and say we were successfully able to get the threshold up in two-thirds of the people that we put through this process. Therefore, we think it's a good thing because now if you ask the right questions at the restaurant and they make a mistake anyway, you probably won't know the difference. On the other hand, being on the therapy significantly increases your risk of having anaphylactic reactions."

When I first interviewed Sicherer, Palforzia was still in clinical trials, and no one knew whether the FDA would approve it. He explained that the controversy in his field, and on display at allergy conferences, was around whether or not oral immunotherapy was a silly thing to do or the best thing to do in light of all the data on the rate of serious reactions.

"On average it seems from the studies that we have so far, that you are going to have more allergic reactions and more anaphylaxis than you would if you avoided it," Sicherer said. "So what the family or individual is worried about is an accidental exposure resulting in a reaction. They're avoiding that by asking a lot of questions and maybe not eating in certain places, or not buying the cookie after all, or whatever it might be. They might have an accident and they might end up with a reaction, but if you submit yourself to this therapy and you're eating some of it every day, people do have reactions out of the blue while they're on their regular dose and they have reactions on their way there. When you just look at the studies, the people who are in the placebo group have less anaphylaxis and allergic reactions than people who are in the treatment group. Is that any better than having to worry about whether a restaurant made a mistake or not? I don't know the answer, but I'll tell you that when I talk about this with a family, there are families that will say, 'Yeah, I don't want to do that,' and other families that say, 'Sign me up yesterday.' I think the experience the families have had in their day-to-day life and what these issues mean to them is what their decision-making comes down to."

In other words, it is a decision based on how each individual or their family thinks about the relative risks of OIT treatment versus avoidance. And that thinking is heavily dependent on what they've gone through in the course of their allergic disease. As Stacey Sturner argued earlier, these are extremely personal choices for patients to make. Not everyone will choose the same path through their treatment.

In their final ICER report, the panel of experts that was asked to review all the available data on Palforzia (called by its scientific name, AR101, here) weighed in against normalizing its use as a first-line treatment. For them, the benefit of two-thirds of study participants being able to tolerate up to 600 milligrams of peanut protein was not enough to counterbalance "a significant increase in gastrointestinal symptoms, systemic allergic reactions, and epinephrine use."[7] Furthermore, the panel felt that neither positive changes in the patients' overall quality of life nor reductions in the number of reactions to accidental exposure to peanuts had been demonstrated.

The panel concluded, "Thus, there is only moderate certainty of a comparable, small, or substantial net health benefit and a small (but non-zero) likelihood of a negative net health benefit for AR101 compared with strict avoidance and rapid use of epinephrine (promising, but inconclusive). Given the need for frequent visits to doctors during the dose escalation phase and the frequent adverse events, it will be important to ensure that patients receive adequate informed consent and that their preferences are carefully elicited prior to initiating desensitization therapy with AR101."[8]

In other words, the ICER panel recommended that clinicians who intended to offer Palforzia make sure that patients understood all the possible outcomes and fully agreed to the treatment. First, that they were likely to experience side effects and that there were risks associated with undergoing the treatment. Second, that not all patients benefited equally from the treatment and that the patients would need to continue maintenance therapy for an indefinite amount of time. Patients would have to decide, along with their doctors, what the best course of action was after being fully informed of all the potential risks and benefits. In the end, however, it remains the patient's decision.

Viewpoint #3: The Company's Perspective

Palforzia was created by a company called Aimmune Therapeutics. Aimmune itself was formed after a Food Allergy Research & Education (FARE) research retreat held in 2011, which was attended by food allergy patients, allergists, research scientists, and representatives from the NIH. The goal of the meeting, as stated on Aimmune's website, was to "shift the focus from funding basic research to finding a food allergy cure by identifying the approach with the greatest likelihood of gaining FDA approval."[9] The result of the meeting? The formation of a new corporation called the Allergen Research Corporation in 2011, which became Aimmune Therapeutics in 2015.

From the beginning, the focus of Aimmune's research was oral immunotherapy. In January 2020, after nearly a decade of research and clinical trials, the FDA approved the use of Palforzia, making it the first federally approved treatment for food allergy. A mere ten months later, Nestlé acquired Aimmune (via its subsidiary Nestlé Health Science) for more than $2.1 billion. The food manufacturer's initial investment in Aimmune ($145 million) occurred in 2016, when Palforzia was still in its early clinical trials. The deal was followed by subsequent investments in Aimmune Therapeutics in 2018 and 2020, totaling $473 million and giving Nestlé a 25.6 percent equity ownership stake in the company even before the acquisition.[10]

Aimmune's code of business conduct and ethics, publicly available on its corporate website, states that it adheres to the highest standards of business ethics. Going further, the code calls for "a higher standard than required by commercial practice or applicable laws, rules or regulations." In other words, Aimmune aims to be a better pharmaceutical company, and one could argue that, from its beginnings in food allergy advocacy, it hasn't been a typical pharmaceutical company. And yet, following the acquisition, it is part of one of the largest global food manufacturers with its own history of not so ethical[11] business practices (like heavily marketing its expensive baby formula to impoverished mothers in developing countries, and arguing in support of its massive bottled-water enterprise that access to water is not a "right" but a "need"). Aimmune was acquired by Nestlé for $34.50 per share, which valued the company at $2.6 billion, a roughly 50 percent increase in Aimmune's share price since 2019. In other words, investment in Aimmune was a good financial move. Why? Because Palforzia, like EpiPen, has a prescription monopoly and will likely be a leader in the OIT market for decades, since it was the first and has a jump on brand name recognition among allergy patients. With the backing of a powerhouse like Nestlé, Aimmune will likely continue to be a leader in food allergy treatments.

At this point, you may be wondering what interest a major food manufacturer might have in a food allergy therapeutics company. When I discovered the connection between Nestlé and Aimmune, I was puzzled at first, too. But as I collected stories about food labeling laws (manufacturers are required by federal law to list all known or possible ingredients) and accidental exposures to allergens, and as I learned about the rising rates of food allergies around the globe, it made more and more intuitive sense to me. Enormous food companies like Nestlé, Cargill, and Archer-Daniels-Midland have a vested interest in making sure that their products have a large market share. If the number of food allergy sufferers continues to grow apace, that is not good for the bottom line. Food packaging has also come under fire from food allergy families for not being easy enough to read for allergens. Someone eating one of its cookies and dying is not good for a company's image or its profit margins. If you are a food company executive, then it is simply good business to support "easy" and safe fixes[12] for food allergies. It protects both the consumers and the shareholders: a win-win for everyone. From the corporate viewpoint, oral immunotherapies are "effective" treatments for allergies because they lessen the risk of corporate liability.

You might argue that I'm being too cynical here. And maybe I am.

But in 2020, Nestlé's food revenue was $76.8 billion. Its investment in Aimmune pales in comparison. If treatments like Palforzia succeed, that means fewer people reacting to Nestlé's products and fewer lawsuits. Both companies stand to gain a lot from Palforzia's continued success.

REAL-WORLD EXAMPLE #2: JAK INHIBITORS
FOR ATOPIC DERMATITIS

Treatments for atopic dermatitis have, until recently, been extremely limited and far less effective in alleviating the worst symptoms of eczema than either patients or their clinicians might desire. Typically, topical corticosteroid creams are prescribed to help control some of the worst symptoms, but their effectiveness is variable and often wanes over time. Their long-term use is also not recommended since patients can develop serious unwanted side effects such as thinning skin or ulcers. What's more, once steroids are discontinued, patients often experience severe flare-ups as their condition rebounds from treatment. As we've already seen, the recent development and approval of Dupixent has given both patients and clinicians a new treatment option—and renewed hope for more effective control of symptoms. However, Dupixent doesn't work for everyone and it, too, can produce unwanted side effects.

When I spoke to clinicians who specialize in treating atopic dermatitis and asked them about new treatment options on the horizon, they frequently mentioned a new class of drugs known collectively as Janus kinase (JAK) inhibitors. The Janus kinases are a group of four enzymes (types of proteins) whose job is basically to add the class of chemicals known as phosphates to other molecules. This addition signals other molecules to activate or become inactive. Think of them like tiny switches for a variety of functions involved in different processes inside your body. In allergic conditions (and in many autoimmune disorders), JAKs are part of the signaling mechanism that helps to activate cytokines, our old inflammatory friends. That means that blocking them can aid in tamping down inflammation caused by a variety of immune responses.

Different JAK inhibitors target different types of Janus kinases and are used to treat different immune-mediated conditions, from rheumatoid arthritis and Crohn's disease to atopic dermatitis. In clinical trials, they have shown great promise in controlling different inflammatory responses. In fact, several different JAK inhibitor drugs have already been approved by the FDA. However, in December 2021, the FDA announced that it would require "black box warning" labels on four JAK inhibitors. Black box labels are the FDA's highest category of risk. If a drug has been shown in clinical safety trials to cause serious or life-threatening side effects, the FDA requires that the manufacturer specify and clearly highlight those dangers on the drug's label. For example, the FDA found that one class of oral JAK inhibitors used to treat arthritis (called Xeljanz) significantly increased the risk of blood clots, cancer, serious heart-related events, such as heart attack and stroke, and death.

This is the scientific context we needed to take a closer look at the first (and of this writing, one of only two) FDA-approved JAK inhibitor for atopic dermatitis, or eczema: ruxolitinib.

In September 2021, the FDA approved a new small-molecule drug called ruxolitinib for topical use in atopic dermatitis, sold under the brand name Opzelura. Opzelura is a topical cream used on affected areas of the skin twice daily. It is for short-term use only, since common adverse reactions can include hives and other infections (bacterial, viral, or fungal). Opzelura is typically prescribed to patients ages twelve and up for mild to moderate atopic dermatitis that is not well controlled with use of steroids or Dupixent.

In phase 3 clinical trials, Opzelura performed well and was well tolerated by a majority of the patients, with 50 percent seeing marked improvement in their conditions as compared to the control group.[13] Patients using the cream reported significant reductions in itch, the main complaint of most eczema patients, within hours of the first application. In these initial trials, none of the patients using Opzelura developed clinically significant adverse reactions at the site of application of the cream.

To use Opzelura or not? As you can see, this is a different decision from the one food allergy sufferers face. People with atopic dermatitis are in no danger of dying from their conditions. That being said, eczema is one of the worst allergic conditions to live with, since it can have huge effects on a person's quality of life. The definition of "effectiveness" under these unique circumstances changes depending on who you're talking to and what aspects of the treatment you're focusing on.

Viewpoint #1: The Patient's Perspective

James Hansen is a hardworking father and husband who lives in Florida and loves to play sports in his spare time. I first "met" James while lurking on Reddit forums related to eczema. I noticed that he was an advocate for the use of biologic drugs over steroids, so he seemed like the perfect person to explain what eczema patients like him think about when they're making treatment decisions. In January 2022, in the midst of yet another wave of COVID-19, we hopped on a Zoom call to discuss his experiences with atopic dermatitis and his recent decision to try Opzelura.

"You caught me on a good day," James said. "My skin feels really good today. Had you called me about four months ago, I would've been completely red, flaky, and miserable. I have made a lot of progress over the last few months, but I've struggled with both allergies and eczema my whole life."

James, who also has food allergies and mild asthma, first developed eczema as an infant. He cannot remember a time when his skin was not an issue in his life. In fact, he candidly tells me that when he was much younger, he had suicidal thoughts because of his condition. In many ways, his allergic conditions have been the focal point for much of his life.

"People don't understand it," he said. "You try to explain that you have a rash or inflammation, and people think you're just whining, like it's not really a problem. But when I have a bad outbreak, I don't want to be seen. I don't want to see anyone. I'm just depressed and hiding."

The triggers for James's outbreaks are . . . everything. Sometimes his skin is aggravated by something topical that he comes into contact with; sometimes it's from his food allergies; other times it's related to stress (and his job is stressful); and as a father with young children (one son and one soon-to-be newborn), sometimes it's just a lack of sleep. For decades, James has mostly tried to manage his worst symptoms—itchy, red, raw, oozing skin—with steroids.

"My thinking had always been, 'What's going to get me feeling and looking good as soon as possible, no matter what?'" James explained. "Whatever the doctor said was going to help, I'm going to try it. And topical steroids were very useful for a while. Then, over time, they just stopped working, so they prescribed me stronger and stronger steroids. Then they stopped working altogether, and I had to go on oral steroids, which cleared up my skin instantly."

But, as we've already seen, steroids have terrible side effects. Neither topical nor oral steroids can be used continuously, and James had already noticed that his skin was thinning as a result of prolonged use. James was trapped in a vicious cycle of steroid use. His skin would get better, but then once he weaned off the oral steroids, his skin would rebound—often getting much worse in the process.

"It was like my skin had become addicted," James said.

James told me that when it comes to treatments, he's tried everything. He's tried holistic remedies, like using oils or Vaseline. He's tried supplements to support the skin. But steroids were the only thing that worked, and even they weren't effective in the way that he needed them to be. In the end, James was using three different types of topical steroid creams: one for his face, one for his body, and one for his scalp. Even on the steroids, he would often scratch at his skin and claw at his face in his sleep, waking up covered in blood. To put it mildly, steroids were not a cure-all for his issues. In fact, he came to believe that they were making his eczema worse overall. That's when he began searching for more effective treatments with fewer negative side effects. In particular, James wanted something that he could use temporarily and then stop using without the danger of another bad flare-up.

"That's when I heard of this newly released drug called Opzelura that is nonsteroidal," James said. "It doesn't do the same damage; it doesn't thin the skin like steroids do. So I figured, let me try this."

When I asked James about the black box warnings, he said he had done a lot of research on the possible side effects of using JAK inhibitors. At first, he was deeply concerned, but then he realized that the warnings were for the oral versions of the drugs, and not the topical applications like Opzelura. James knows that a new drug doesn't have a long-term safety profile yet, so there will always be risks. But for him, the possible benefits far outweighed those risks.

"Life is not good when your skin is really bad," James explained. "I am miserable when my skin is in full flare. Sometimes it just feels like those risks are worth it. I also looked at the percentages. What percentage of the people in the study had a negative impact? If it's less than ten percent, to me, that's a great risk. That's a ninety percent chance that I'm not going to have that issue."

When we spoke, James was a few months into his new combined treatment with Dupixent, which he takes daily, and occasional spot treatments of Opzelura, which he applies whenever he has a flare. His skin looks terrific. On our Zoom call, with his small son playing in the background, James said he is happy and feels "normal" again. He's able to sleep and has energy for work, his wife, and their son. And soon, James and his wife will have a new baby. Opzelura, James attests, has changed his life for the better. As for those risks? He's going to try to wean himself off Dupixent and keep Opzelura for future flare-ups. Ideally, a fully effective treatment for James would be something that cured his eczema—and maybe his food allergy and asthma, too. But for now, Opzelura is close enough.

Viewpoint #2: The Expert's Perspective

When I asked Dr. Peter Lio, a clinical assistant professor of Dermatology and Pediatrics at Northwestern University Feinberg School of

Medicine and the founding director of the Chicago Integrative Eczema Center, how he thinks about "effectiveness" in relationship to treatments, he responded with enthusiasm for the question itself. It's something that he has thought a lot about over the years as he's dealt with severe atopic dermatitis patients, many of whom have recalcitrant eczema and have struggled for years to find better options for managing their condition.

"This is a really good, really deep question that can take us down some rabbit holes," he said, "despite the fact that it seems so straightforward."

Echoing his colleagues who work in food allergy, Lio stressed that the definition of an effective treatment relies almost wholly on the patient's perspective. It's impossible, from the outside, to guess how someone is going to feel about their skin. It's not always as simple as looking at their skin and assessing their lesions.

"I have some folks who look pretty bad," Lio explained, "but are happy and do not want to change things at all. On the other hand, I have super severe folks who we've been able to get almost clear almost all the time, but they are still deeply unhappy and want more. Both of these positions are valid, and a big part of our job is going through the shared decision-making process, weighing where they are at, where they came from, and potentially where they could be given the state of the art."

While things like the size or color or condition of a skin lesion can be measured and independently assessed by a clinician, things like what that skin lesion feels like or its effect on the patient's day-to-day life cannot. As Lio and other clinicians are quick to point out, using clinical signs (such as scoring on the Eczema Area and Severity Index—a scale used in clinical trials to assess the severity of atopic dermatitis) is not enough to decide the overall effectiveness of any allergy treatment.

"We have to balance efficacy, safety, tolerability, and accessibility for each and every medication," Lio said. "And of course one person's scary side effect is totally reasonable to another person."

For someone like James, the possibility of increasing his future risk of a heart attack or cancer seems minimal in relationship to how his skin looks and feels today. James is also relatively young and otherwise healthy, so he might weigh those risks differently than someone who has underlying conditions. Others may look at a black box label on a new drug like a JAK inhibitor cream and decide that they are not comfortable with those risks no matter how statistically small. For other patients, especially those without good healthcare coverage, the cost alone might be the deciding factor. At current prices, a 60-gram tube of Opzelura costs $2,013. (But with good insurance, the amount that a patient owes may be as low as $10 per tube.)

To help patients navigate their treatment options, Lio recommends that clinicians use a relatively new diagnostic survey called the Atopic Dermatitis Control Tool (ADCT). It asks patients to answer six questions about their experiences over the course of the previous week. The questions touch on issues like the quality of their sleep, how much their skin condition impacted their daily activities and mood, and the severity of their itching. The point total helps both patients and their doctors track their progress to see if their current treatments are effective in terms of improving their overall well-being and quality of life. It helps take the arbitrariness out of the decision-making process. Using such a tool, Lio is more confident that he can tell whether or not a JAK inhibitor like Opzelura is truly working to address his patients' most serious symptoms and concerns.

"I think this approach is the future," Lio suggested, "and it's a big step forward."

Viewpoint #3: The Company's Perspective

Opzelura is manufactured by Incyte, a biotech company. Formed in a merger in 2002, Incyte focuses on the discovery and development of new biologic drugs for use in oncology and dermatology. In other words, the company attempts to harness the power of our growing body of scientific knowledge about the immune system to treat a variety of immune-mediated medical conditions.

Ruxolitinib, the small-molecule drug used in Opzelura topical cream for atopic dermatitis, was first approved by the FDA in 2011 for use in treating myelofibrosis, a rare type of bone marrow cancer. The oral version of ruxolitinib, called Jakafi, blocks both JAK1 and JAK2 and can cause serious side effects like blood clots and cardiac events detailed earlier in the chapter. Typical side effects of the topical form of ruxolitinib are diarrhea, bronchitis, elevated eosinophil counts, runny nose, and hives. Ruxolitinib is a profitable drug for Incyte, pulling in 70 percent of the company's total third-quarter revenue in 2021 with $547 million in sales. Its profit margin showed growth of more than 12 percent from the same period in 2020. Any pharmaceutical company would be overjoyed with that year-on-year growth, and those numbers don't even include sales from the more recently approved Opzelura. As we now know well, the number of allergy patients shows no signs of diminishing, so Incyte stands to make a healthy profit from its line of JAK inhibitors for atopic dermatitis. When I originally interviewed patients and clinicians, Opzelura was also the only FDA-approved drug, so Incyte was enjoying a brief market monopoly as other pharmaceutical companies with JAK inhibitors made their way through clinical trials. With all that in mind, Wall Street analysts expect the drug to generate anywhere between $600 million and $1.5 billion per year for Incyte by 2030.

The Institute for Clinical and Economic Review, however, has a few concerns about the new biologic drugs, both monoclonal antibodies like Dupixent and JAK inhibitors like Opzelura. For one, the independent panel of experts pointed out that both classes of drugs will only be prescribed and covered by insurance once patients can show a history of other, unsuccessful treatment regimens. Opzelura is specifically indicated for use in those with mild to moderate atopic dermatitis, but as its ICER report[14] points out (and as we've already learned), there are currently no good guidelines or standards for determining exactly who falls under this category. For instance, James clearly had severe bouts of eczema, so technically he wouldn't. (Clinicians often have to use false diagnosis codes to obtain insurance coverage for their patients. This practice is widespread, and not just in relationship to allergy.) To use these new drugs successfully, ICER argues that diagnostic tools need to be improved and standardized. A worthy goal, to be sure, but one that will be difficult to enact given the disjointed nature of allergy care.

ICER also flagged safety concerns and recommended that patients not be given the treatment for extended usage until more data from safety trials had been collected. While the drugs seem promising, the reality is that we know relatively little about their overall effect on immune function over longer courses of treatment.

As is often the case with newer federally approved drugs, the patients who use them in earlier stages are part of a large uncontrolled experiment. Pharmaceutical therapies for atopic dermatitis are often used in tandem with one another, muddying the scientific waters. It can be hard to tell, as James admitted, if Opzelura alone was making the difference for him, or if it was really the combination of Dupixent and Opzelura. But for many patients like James, being a part of real-world data on new drugs is preferable when they've run out of alternatives.

From a corporate perspective, monoclonal antibodies and JAK inhibitors fill a needed gap in skin treatments while the companies that manufacture them stand to make a great deal of profit. For them, effectiveness is always seen from both a medical perspective (Does the drug improve patient well-being on clinical assessment scales?) and a financial calculus (Are the drug sales healthy and is the patient population robust enough to ensure continued growth?). From both of these viewpoints, Opzelura looks like a clear winner.

WHAT EFFECTIVENESS REALLY IS, REDUX

Our understanding of the human immune system is still developing; many facets of how our bodies interact with the world around them remain shrouded in complexity and uncertainty. Yet in the last decade, great strides in basic immunology have led to promising innovations. New allergy treatments like Palforzia and Opzelura demand that patients and their doctors carefully weigh the potential risks and benefits. Because they are attempts to tinker with our immune functions, all of these treatments will likely come with side effects.

James Hansen often repeats the tale of his treatment journey on subreddits devoted to topics related to eczema and topical steroid withdrawal. His experience might matter to other patients just as much, or perhaps more, as scientific data on Opzelura's effectiveness from a more clinical standpoint. Individuals who are desperate to find more effective treatments for their condition can also relate to James's personal story in a way that they cannot relate to hard facts. In his story and those of others like them, they discover not only new treatment options, but hope.

This book is no different. It's filled to the brim with stories. Patient stories, like James's and Stacey Sturner's, and allergist stories, like Dr. Sicherer's and Dr. Lio's. These stories are influencing you, the reader, right now.

Imagine again, your child with severe peanut allergy. Now would you undergo treatment with Palforzia, or would you elect to keep avoiding the allergen? If your child balked at eating the food with the peanut protein, would you force them to continue? What if they developed stomach pain?

What did you decide? How do you think you would feel about your decision? Confident? Guilty? Hopeful? Anxious? All of the above?

These are the dilemmas patients and caretakers of patients with moderate to severe respiratory, skin, and food allergies face on a regular basis. Their need for better information, more access to quality time with expert healthcare professionals, and financial and emotional support throughout the treatment process is growing at the same pace as allergy rates. The only questions left to ponder are: What are we as a society going to do about all this? What would an effective social or collective response to the increase in allergies even look like?

CHAPTER 10

Allergy Is a Social Problem, Too

———————

THE IMAGE OF ALLERGY SUFFERERS IN AMERICAN CULTURE

When I was thirteen, I loved *The Goonies*. Maybe you've seen it: a 1980s Steven Spielberg movie about a bunch of preteens and teenagers—the "Goonies" of the title—and their attempt to decipher a 250-year-old pirate treasure map to save their working-class homes from the clutches of real estate developers who want to build a golf course on the land. Its central hero, Mikey Walsh, is a young lad with asthma. The first time we see Mikey, he uses his inhaler as his older brother lifts weights and calls Mikey a "wuss."

"I'm not a wuss!" Mikey yells back.

At the start of the film, Mikey's mom admonishes his older brother to make sure that Mikey doesn't set one foot outdoors, due to his medical condition. "If he's coming down with asthma, I don't want him out in the rain," she says.

As soon as their mom leaves, Mikey's brother says, "You want a breathing problem? You got one."

Then he grabs Mikey, wrangles him into a classic eighties movie headlock, and proceeds to give Mikey a "noogie."

Throughout the movie (which, incidentally, held up quite well when I rewatched it recently), we see Mikey use his inhaler frequently. It serves the same function in each scene: as punctuation to Mikey's visible nervousness, anxiety, or fear. In fact, he uses the inhaler so frequently that the allergy researcher in me grew concerned. The rate at which Mikey puffs would be detrimental in real life—no one can handle that many doses of inhaled steroids or bronchodilators in quick succession. Mikey, played by a very young Sean Astin of *Lord of the Rings* fame, is small and slight. He wears braces. Although he is brave, the de facto leader of the Goonies, he is also depicted as a dreamer.

By the end of the movie (spoiler alert), Mikey and his pals are victorious; they have outwitted some nefarious folks to save their families' homes. The pretty girlfriend of his older brother approaches Mikey and says, referencing an earlier kiss, "The parts of you that don't work so good are gonna catch up to the parts that do." In response, Mikey throws his inhaler over his shoulder, muttering, "Oh, who needs it?"

The message here is clear. Having faced his fears, Mikey is no longer "weak" or asthmatic. He doesn't need coddling anymore; his courage has cured him. Asthma, we learn, is for wusses, and Mikey is no wuss.

I am bringing up this delightful children's movie here to make a larger point: The media—often unconsciously—molds our image of the typical allergy sufferer. Growing up in the late 1970s and early 1980s, I learned, via cultural imagery like *The Goonies,* that having a respiratory allergy was a limitation. Beyond that, it was a physiological hamstring and indicated that someone was either more vulnerable (at best) or nerdier (at worst). Many films, TV shows, and novels, both then and now, associate having a respiratory or food allergy with being "a loser," "a geek," or a cultural underdog. Or else allergies are used as a plot point, for comic relief or as an easy backstory.

Milhouse, a "nerdy" character in the long-running TV show *The Simpsons,* is famously allergic to wheat, dairy, and his own tears. In the movie *Monster-in-Law,* Jennifer Lopez's character is purposefully exposed to almonds by her soon-to-be mother-in-law. The entire scene is played for laughs: Lopez's character immediately starts coughing and says her tongue feels funny; her face swells hugely. Comedian Louis C.K. jokes about peanut allergy sufferers being evolutionarily challenged and suggests that "maybe if touching a nut can kill you, you are supposed to die." Culture primes us to think in certain patterns about a subject. In the case of allergy, these portrayals—while arguably harmless on the surface—can have lasting consequences. A case in point: food allergy bullying and the film *Peter Rabbit*.

In this 2018 movie based loosely on the famous children's tale, a band of rabbits (led by Peter, of course) is in a battle with old farmer McGregor's nephew, Thomas. McGregor has died of a heart attack, and Peter and his friends have taken over the garden entirely, when Thomas comes to reclaim the estate. A battle for the garden ensues. In a scene that led many food allergy advocates to protest the film, Peter and his friends pelt Thomas with various fruits. As one rabbit picks up a blackberry, he realizes that Thomas is allergic to them. The rabbits begin to aim the blackberries at Thomas's face, and one flies into his open mouth. Thomas swallows and immediately begins reacting. He reaches into his pants and pulls out an EpiPen, stabbing himself in the upper thigh and falling back. The rabbits think they've defeated him until the adrenaline makes Thomas rise back up, prompting Peter to exclaim, "This guy's like some kind of sorcerer."

People were upset about this scene for several reasons, not least of which was the fact that it took place in a film aimed at younger children and based on a children's story. What, exactly, was the message of this scene supposed to be? That it was okay to throw allergenic food at someone with a severe food allergy? That doing so would be fine as long as the person had an auto-injector?

On Twitter and other social medial sites, #boycottpeterrabbit started to trend. Under social pressure, Sony issued a statement saying that the studio regretted making light of a serious medical condition, and not being "more aware and sensitive" to the plight of those families coping with severe food allergies. But for many, the damage was already done; millions of children had viewed the offending scene. Some parents and advocates expressed worry that seeing food bullying as acceptable behavior onscreen might make it seem more socially palatable offscreen.

Allergy bullying is a real phenomenon, especially in school settings, such as lunchrooms and on playgrounds. I've heard countless stories like that of a woman I'll call Jaime. Jaime grew up with severe eczema. Her skin began reacting to things while she was in kindergarten. Over the years, it would get better, flare up a little, and then be nonexistent for a while, her skin returning to normal. In 1982, when Jaime was in the fifth grade, her skin erupted once again. This time the reaction was serious and sustained.

"I would scratch so badly that I had huge open wounds on my hands and my arms," Jaime remembers. "I would put gloves on at night. I would scratch through the gloves. I'd wake up in the morning and the gloves would be stuck to my hands from the weeping cuts." Jaime pauses and her voice goes down a register. "Then the teasing started."

Jack, a boy in Jaime's grade who also lived in her neighborhood, noticed her skin. On the bus ride to school, Jack would tease Jaime and encourage other boys on their bus to make fun of her "alligator skin." Jack taunted Jaime throughout grade school and middle school.

"It affected me emotionally for so long," Jaime recalls. "I don't know if he ever realized how damaging what he did to me was. I don't think he did. My parents went to his parents to complain, but his parents just said boys will be boys. Nothing was ever done. It had a huge impact on me."

The unpleasant memories of that time in her life still linger. The trauma of being teased mercilessly lasted long after her skin cleared up, so much so that she thinks of it as a part of the foundation of her entire personality. For the children who report being stigmatized or bullied about their allergies, the experiences often leave permanent social scars.

When I talk to young adults who have an allergic condition (and as a college professor, I come across a lot of them), I always ask about experiences of teasing, bullying, or feelings of social ostracism in relationship to their conditions. Happily, most tell me that they didn't really experience any negativity around their condition from close friends or family throughout grade school and high school, and still don't. That being said, almost in the same breath, they'll often admit that they don't like to bring their inhalers or adrenaline auto-injectors with them to social outings or make too much of a fuss about their conditions when they are hanging out with their friends. They prefer, if they can, to "blend in" with their nonallergic peers. Their motto seems to be: Do not disrupt normal social events with allergy needs.

Dr. Eyal Shemesh, a psychiatrist who works with patients at the Elliot and Roslyn Jaffe Food Allergy Institute at Mount Sinai, is not surprised by reluctance to carry prescriptions. It is, he told me, part of an avoidance strategy that younger allergy patients often develop during the early years after their diagnoses. No one wants to be reminded at every turn that they are different from their peers, or that they may die from their condition. It's scary and it is often easier not to think about it. It also protects them from any social pushback they might experience if they were openly and visibly allergic. There is a difference, though, between the negative stigma, undergirded by cultural images, that comes with having an allergy, and blatant allergy bullying.

"Bullying is a very specific construct," Shemesh told me. "It is a repeated pattern that's aiming at hurting the recipient."

As the movie *Peter Rabbit* showed, the goal of bullying is injury. Shemesh was surprised by how prevalent bullying was among food allergy patients. It is easy to do a Google or social media search and find instances of food allergy bullying: a twelve-year-old girl with dairy allergy who had nacho cheese dip rubbed in her face; a thirteen-year-old boy who died after having a slice of cheese thrown on his bare skin; a young adult male recalling how many times someone used a peanut butter sandwich to shoo him away from a lunch table. One 2011 study found that children and adolescents with asthma were also likely to experience bullying due to their condition.[1]

"What is more interesting is that in many cases, the parents don't know," Shemesh said. Even when directly questioned by their parents, children with allergic conditions conceal their negative experiences at school or at social events. Only when a neutral party like their clinicians asked them about their social interactions in relationship to their allergy did they cop to having been bullied. In a 2012 study, Shemesh found that more than one in three children have experienced bullying related to allergy.[2] But in another study of food allergy parents, only one in five reported that their family had experienced bullying.[3] In other words, parents don't necessarily know when their allergic kids are having social problems.

Shemesh sees allergy bullying as a larger social and cultural problem, one that we need to do much more to solve. In the United States, children are often counseled to simply ignore bullies. When parents do find out about bullying, Shemesh said that they usually try to solve it themselves, often by talking to the parents of the offending children. Both approaches are ineffective at best, Shemesh argued.

"It's not the child's issue," Shemesh said. "We need to work together to stop it."

And that is the crux of the issue here: We need to work together to help the growing number of children and adults with allergies. This chapter explores our cultural attitudes about allergies, our empathy for strangers with irritated immune systems, and what that might mean for the creation of allergy and other environmental policies in the future. Recent public debates over serving peanuts on airplanes, the creation of allergy lunch tables and allergen-free spaces, food labeling laws, allergy bullying, and depictions of allergies in movies and on TV shine a spotlight on how people think about social responsibility in relationship to allergy prevention and care.

The problem of allergies highlights all the ways in which we are interconnected and ultimately indebted to one another for our health and well-being. If allergy is caused by everything we're doing, then it will take all of us to solve it.

WHAT AMERICANS THINK ABOUT ALLERGY

In 2019, a few months before the COVID-19 pandemic began, I conducted a survey of one thousand Americans[4] to try to better understand some of our cultural attitudes and beliefs about allergies. The results represent the viewpoints of every conceivable demographic and include both people who have allergies (56 percent)[5] and those who are blissfully allergy-free (44 percent). When I designed the survey questions, I had just begun to interview experts and patients for this book. I based my questions on the historical research I had already done and various media representations of allergy sufferers, not on the conversations I would have throughout the next three years. As an anthropologist, I suspected that our long history of associating allergies with neuroticism, women, city dwellers, and the highly educated would have consequences for how we view modern-day allergy sufferers. I expected to discover that, overall, Americans thought of people with allergies as somehow "weaker"—either physically, emotionally, or both—than their nonallergic peers. But what I found surprised me.

The results of my survey suggest that the majority of Americans[6] do not view people who have allergies as weaker than nonallergic people. Only one in four respondents indicated that they felt those without allergies were physically stronger, and just 14 percent of respondents felt that people without allergies were emotionally stronger. These findings suggest that our collective experience with allergies is beginning to shift the cultural narrative. (As a case in point, remember back to our exploration of the nineteenth- and early-twentieth-century viewpoint that people with allergies were somehow weaker and more neurotic—and that their neuroticism was actually causing their disorders.)

When I asked respondents if they felt that the parents of children with either food allergies or asthma were overprotective of their children, most said no (59 percent and 69 percent respectively). That being said, about 39 percent of people did agree that parents of kids with food allergy were too worried about them (and men were more likely than women to think so). And at least 30 percent of people agree that parents of those with severe respiratory allergies or asthma were too anxious about their kids' well-being. This suggests that while Americans do not believe that children with allergies are any weaker, some of us do think that parents may be overreacting to their children's conditions.

In interviews and conversations with nonallergic people, I often noted a mild suspicion that those with allergies or their caretakers might be exaggerating their symptoms or situation. Survey results bore this out, too. Most Americans indicated that they personally know at least one person with allergies (72 percent); just over 35 percent of survey respondents reported that they thought people with allergies "sometimes" exaggerated their symptoms. Around the same percentage (41 percent) said that they had personally suspected that someone they knew was "faking it" or lying about their allergies. Interestingly, younger people (ages eighteen to twenty-nine) were almost *twice as likely to be suspicious* than older (over sixty) respondents. That being said, most respondents don't think we're being too accommodating to those with allergies; only 36 percent of Americans argued that we are being over-vigilant (and most of those think we're only "somewhat" too accommodating). Again, those ages eighteen to twenty-nine were far more likely than older Americans to think that schools, restaurants, airlines, and other institutions were going "too far" to accommodate allergy patients.

Younger Americans are far more likely to have grown up with allergic classmates, friends, or family members. So why were they more apt to suspect them of exaggeration and less likely to want society to accommodate their needs? Looking over the results, I started to wonder if more familiarity with allergies might breed some form of contempt, or at least less empathy. Perhaps their direct experiences with allergies have normalized the condition to the extent that allergies are now seen by the youngest generations as part of "normal" life, and thus not particularly deserving of special treatment.

While 48 percent of Americans think allergies are getting worse, 67 percent think that more people have allergies than they did twenty years ago, and 81 percent believe that having allergies negatively affects a person's quality of life, allergies still rank as some of the least sympathetic illnesses. When I asked people to rank eight common diseases from "most to least sympathetic," hay fever/respiratory allergies were by far the least likely to conjure up compassion or commiseration. And food allergy patients were the second least likely to be seen as sympathetic. Those with severe eczema came in at third least likely to generate empathy.

Which diseases were seen as much more serious—and thus deserving of more caring? Heart disease, chronic pain, and skin cancer.[7] Americans older than forty-five were far more likely to think that heart disease was the worst condition, whereas younger Americans (eighteen to twenty-nine) were more likely to think skin cancer was more deserving of empathy and concern. This tracks with their stages in life. If you are older than forty-five, you're more likely to be afraid of heart disease or have experienced chronic pain. Generally, we tend to sympathize more with the conditions that we can imagine ourselves having, and those that are more likely to kill us.

I was pleasantly surprised by the majority of these findings, yet they still leave a lot to be desired. On the positive side, most Americans are on the same page, believing that allergies are worsening, more people are getting them, and having them negatively affects quality of life. On the less positive side, while most Americans are willing to make accommodations for allergy sufferers, they are clearly not always empathetic to patients' lived experiences. Regrettably, I was not surprised to discover that most Americans feel as though allergy patients fib at least sometimes about the severity of their allergies. On reflection, our American attitude toward allergy is a decidedly mixed bag. But if these survey results are correct, and if Gen Z attitudes change as they begin having children of their own (as they likely will), then it may be slowly shifting . . . and for the better.

But will all this increased awareness and empathy for allergy sufferers translate into better policies at the local and national levels and for society as a whole in the future?

A BRIEF LOOK AT ALLERGY POLICIES, REGULATIONS, AND LAWS

In 2015, as I was boarding a flight for Colorado, a young woman in the row behind me began asking those around her to refrain from eating anything that contained tree nuts while on the flight. If anyone had such a snack, she generously offered to pay for a different one. The man next to her said he had a granola bar but wouldn't open it. An elderly man across the aisle commiserated, telling the young woman he had a grandson with the same affliction.

Soon after, the flight crew made an announcement asking all passengers to refrain from eating anything that contained nuts for the duration of the flight and informing us that there wouldn't be any nuts offered as part of the in-flight food service. In front of us, there were a few audible groans of disapproval. Around us, silence. I glanced behind me to see the young woman quietly settling into her seat, a flush of red on her cheeks.

Decisions about whether or not to offer nuts (or other allergenic foods) on planes are not officially regulated by the Department of Transportation. They are left to the discretion of each airline. In this moral and ethical gray zone, in the absence of any formal laws or regulations, most airlines have instituted allergy-friendly policies to help protect allergic passengers. For example, Southwest Airlines, United Airlines, and Air Canada have stopped serving peanuts altogether, even when no allergic passengers are on board, despite the fact that anaphylactic reactions due to inhaled or residual peanut dust remain exceedingly rare. Other airlines accommodate passengers with allergies, but only with advance notice.

Yet even seemingly small concessions like these can create a swift social response. When Southwest Airlines banned all peanuts on flights in 2018, social media sites like Twitter, Facebook, and Reddit blew up with comments, both in support of the measure and critical of it. Most people applauded the move. Others decried how "soft" Americans had become and how unfair it was for a minority of people to affect the habits of the majority. A few commenters even vowed to keep eating nuts on planes despite the ban.

Years later, I often find myself thinking back to that flight. If any of us who were seated in the rows around that severely allergic young woman had not heeded her plea, then we may very well have endangered her well-being. If someone had ignored the temporary ban and whipped open a packet of mixed nuts anyway, even if her allergies had not been triggered, that young woman's sense of security and safety would have plummeted as her anxiety spiked. In that scenario, we—her fellow passengers and human beings—would have failed her. And while nothing of note happened on our flight that day—we all refrained, and the young woman deplaned without incident—I know that other allergy patients who need to rely on the public for their safety have not been so lucky.

In July 2018, Kellie Travers-Stafford's fifteen-year-old daughter Alexi spotted an open container of Chips Ahoy! cookies while visiting a friend's house. Alexi, who had a severe allergy to peanut, thought she recognized the same red packaging as the nut-free version she regularly consumed at home. Feeling confident, she popped one into her mouth.

Alexi immediately recognized the first signs of an impending ana-phylactic response—a tingling sensation in her mouth—and quickly returned home. While waiting for paramedics to arrive, Alexi's mother, Kellie, administered two EpiPens, delivering two full doses of epineph-rine in rapid succession, hoping they might halt Alexi's allergic reac-tion long enough to get her the medical care she now so desperately needed. But despite understanding her food allergy, being cautious about what she ate, and having access to an EpiPen, Alexi was dead within ninety minutes of eating a single cookie that contained peanut.

"As a mother who diligently taught her the ropes of what was okay to ingest and what was not," Kellie wrote, "I feel lost and angry be-cause she knew her limits and was aware of familiar packaging, she knew what 'safe' was."

In an emotional Facebook post detailing the events leading up to her daughter's death, Kellie pleaded for food manufacturers to be more consistent with their labels and packaging. Kellie's goal, she said, was to prevent another family from going through a similar tragedy. By the time I sat down to read her tale, a mere two weeks after her daughter's untimely death, Kellie's post had garnered more than twenty thousand comments and been shared more than 79,0000 times. Various news and media outlets had picked up the story, and Alexi had become part of a growing national conversation about allergies in America and what we should do about them.

. While most reporting, responses, and online comments reacting to Kellie's story were sympathetic, some expressed doubt about whether Nabisco, the company that manufactures Chips Ahoy!, could, or should, be held at least partially accountable for Alexi's death. A representative of Nabisco's parent company, Mondelēz International, responded to events by stating that their corporation takes allergies very seriously and makes every effort to ensure their foods are clearly labeled, adding: "We always encourage consumers to read the packaging labeling when purchasing and consuming any of our products for information about product ingredients, including presence of allergens. (For added reference, the packaging for Chips Ahoy! with Reese's Peanut Butter Cups prominently indicates, on both the front and side panels, the presence of peanut butter cups through both words and visuals.)"[8]

As news of Alexi's death continued to circulate and reverberate, the dividing line between, on the one hand, those calling for increased vigilance on the part of allergy patients and, on the other hand, those calling for stricter labeling on the part of food companies seemed to be the level of allergy awareness. The closer people were to someone with a deadly allergy, the more likely they were to agree with Kellie's entreaty for better packaging. In this case, familiarity didn't breed contempt—it produced understanding, empathy, and outrage.

When talking to allergy experts, I was reminded at every turn that allergy is a community issue. It can be easy to dismiss allergy as a minor medical condition because it so rarely kills anyone. And since allergy is unique to each person and not contagious, it is usually seen as an individual's personal medical problem. But over the course of researching and writing this book, I've come to think of allergy as not just an individual, biological issue, but as a deeply social one.

Allergy sufferers are the first victims of our collective transformation of our environments and daily habits. They cannot possibly hope to avoid coming into contact with allergens—chemicals, pollen, proteins—without the cooperation of everyone around them. Simple actions such as abstaining from eating certain foods in close quarters (for example, on airplanes or at school lunch tables) have become cultural battlegrounds. From bans on snacks containing peanuts on planes to new food labeling laws, each policy enacted to help allergy patients has been met with a fair amount of public pushback. There is a tension, always at play in medical conditions caused or triggered by environmental factors, between individual rights and responsibilities, and the need to protect and promote the health of the community at large. (More recent debates over masking, school and business closures, and social distancing throughout the global COVID-19 pandemic have starkly underlined this.)

Yet what both of these stories highlight is the fact that at its heart, allergy forces us to ask some uncomfortable questions about our duty to one another as members of a society:

Who is responsible for the health and well-being of someone with allergies?

Who is obligated to help keep allergens at bay?

Is it fair to ban certain foods or perfumes or certain trees in public places because they're not good for everyone's health?

How responsible should corporations be for our overall health and well-being?

Should we have rules or laws that limit some of our individual rights in order to protect the overall health of everyone in our communities?

These critical questions reflect what is at stake in crafting social and environmental policies, not just for allergy patients but for overall human health and well-being. Let's take a closer look at how some of these questions play out in relationship to more recent federal and local regulations aimed at protecting allergy patients.

The Regulation of Social Changes:
Food Labeling Laws and Food Allergy

If you grew up before 1990, you may recall that when you went to buy packaged food, there was no way to tell how many calories it contained. An ingredient list had been mandated by the federal government as early as 1906 to help curb the proliferation of unsafe food additives and false advertising, but it wasn't until 1990, as obesity rates began to rise at an alarming rate (along with the number of Americans with a serious chronic illness related to their diet, such as type 2 diabetes), that Congress passed the Nutrition Labeling and Education Act (NLEA), standardizing the use of nutrition labels on packaged food.

At the same time, food allergy rates had begun their own steady climb. The new nutrition labels provided consumers with more information about the foods they were eating, but they did not address the particular issues food allergy patients faced in the supermarket aisles. Dr. Scott Sicherer of the Elliot and Roslyn Jaffe Food Allergy Institute remembers the situation well, in part because his own research would be instrumental in changing it.

"In those days, labels would say things like 'natural flavors' and you wouldn't know what that meant," Sicherer said. "It didn't have to say there was milk in the product because milk could have been a natural flavor. You could have just about anything in the food as a secret ingredient. And then they were also using chemical names. So you'd have to know that 'casein' was a word for milk protein, because that's all it would say."

If you didn't know what casein was and you had a child with a milk allergy, then you were probably in big trouble. And that's exactly what Sicherer and his fellow researchers found in a 2002 study of parents of children with food allergy. Only 7 percent of parents could correctly identify milk as it was listed on fourteen ingredient labels of products containing it. Just over half of parents could identify peanuts as listed on all five product labels that contained it.[9] These are not terrific numbers. As just one example, I doubt that many of us would recognize *Arachis hypogaea* as "peanut" on a granola bar label.

It was clear to Sicherer, parents of young children with allergy, and most practicing allergists and pediatricians that something had to be done to help people avoid their allergens. The largest food allergy advocacy group at the time, the Food Allergy Network (FAN), decided to help track the problem. By the early 2000s, it was receiving hundreds of credible reports per year from its members about packaged foods that had been inadequately—and thus dangerously—labeled.

In a random review of foods completed nearly a full decade after the passage of the NLEA, the FDA found that at least 25 percent were mislabeled or inadequately labeled, failing to list egg and peanuts as ingredients. Food recalls due to unlabeled allergens had skyrocketed. And, as Sicherer pointed out, even when allergens were listed on food labels, their scientific names were often used—not the common names most consumers know well. At the turn of this century, it was becoming all too clear, even to the FDA itself, that the food labeling laws were grossly inadequate.

Enter one of the only federal laws ever to be enacted that touches directly upon allergy: the Food Allergen Labeling and Consumer Protection Act (FALCPA) of 2004.

As a corrective to the NLEA, the law was meant to help consumers discern the ingredients in packaged foods. After the FALCPA, manufacturers were required to list all ingredients that may be contained in any food product, using the common names for all major allergens. The act specified that even trace amounts of the eight most common allergens must be clearly labeled on any commercially available foodstuff.

While this new labeling law has made it easier for allergic people to navigate the supermarket aisle, the system is far from perfect—it tragically failed Kellie Travers-Stafford's daughter Alexi. The biggest problem with the FALCPA is that many allergen-free foods are manufactured using machines and production lines that have been used to make foods that *do* contain those allergens. If Mondelēz International uses the same facilities or machines to make Oreo cookies as they do Chips Ahoy! with Reese's Peanut Butter Cups, then people with food allergies are at risk of cross-contamination. Typically, the manufacturer places a warning or advisory label on foodstuffs that may have been affected in this way.

But, as Sarah Besnoff explained in the *University of Pennsylvania Law Review* in 2014, the FALCPA "contains no specifications regarding how to list cross-contact warnings, no requirements about how food producers should measure cross-contact or report any discovered risk of cross-contact, and no limitations on what a company can include in its advisory labeling. As such, a cursory perusal of any neighborhood grocery will show a wide variety of warning labels, from no warning to 'May Contain . . . ' to 'This product was processed on machinery . . . ' to 'We are unable to guarantee. . . .' None of these warnings explain how this risk of cross-contact was measured (if at all), where in the production process this potential contamination may have occurred, or whether this perceived risk of cross-contact is the result of testing, speculation, or, worse, a nervous legal department."[10]

Most major food manufacturers have now recognized the growing problem of food allergy and have begun opening allergen-free manufacturing facilities in order to guarantee that their products are safe for consumption. It is a step in the right direction and one that takes some of the burden off individual allergy patients to ensure their own safety. That being said, precautionary allergy labeling practices remain unstandardized and unregulated, both nationally and globally. In the continued absence of legal regulation of standards, or even guidance by the FDA, food manufacturers have been left to their own devices. A recent study of American and Canadian food allergy patients found that nearly half assumed advisory labeling was required by law.[11] A third thought the labeling reflected the amount of allergen present in the products. Those who had a history of severe reactions were more likely to completely avoid anything with an advisory label.

In other words, although manufacturers are starting to do better, food allergy patients are still largely left stranded in the supermarket aisle with unclear labeling and incomplete information. The onus for protecting food allergy patients remains almost squarely upon their own shoulders. In the absence of federally regulated rules and policies, food allergy patients and their families have to try to educate themselves about which foods and manufacturers are "safer" than others. As a society, we should be asking ourselves if this is fair or desirable. My guess is that even if we do not have a food allergy ourselves and cannot directly relate to the uncertainty that food advisory labels create for patients, we can all agree that food should be safe for *everyone* to consume. The question going forward is: How should we regulate the food industry so that we can ensure that it is?

The Regulation of Environmental Changes:
Landscaping and Respiratory Allergies

Mary Ellen Taylor has owned her own landscaping business since 1986. For nearly forty years, she's been mulching, mowing, trimming, pruning, and planting near her home in Delaware. She has a small crew that works for her, including her husband, and loves her work.

Fun fact that's pertinent to this story: Mary Ellen has asthma and respiratory allergies herself. As a kid, she was allergy-free. But over the years, she developed them—likely because of her repeated contact with allergens as part of her work as a landscaper (a fate she shares with botanists, as we saw in chapter 5). Her first respiratory allergic reaction was around age thirty, after spreading five yards of mulch in her own yard by herself, and she's found herself in the emergency room for asthma attacks a few times since then. For years, she used a rescue inhaler, a steroid inhaler, and daily antihistamines to keep everything under control. But now she's dialed back on the steroids since she knows that they can negatively affect her bone density and teeth.

"My biggest triggers are cat dander, mold, and grass," Mary Ellen tells me on a phone call in late 2021. I had called her to get a professional landscaper's take on dealing with respiratory allergies and to ask questions about how much pollen factors into garden design and planting. "I'm around mulch, which contains mold, and I'm around grass all the time. So, when I found that out, I was like, oh, great."

For the most part, Mary Ellen lets other members of her crew deal with the mulch now. And although she takes her fair share of soil samples for testing—a typical task for a landscaper—and does some planting herself, she makes sure that she washes her hands. She also never goes anywhere without her inhaler. If she's forgotten it at home, she turns the car right around.

I ask Mary Ellen how her allergic asthma diagnosis affected her job, as someone who deals with plants for a living.

"I just became really focused on where I was in a landscape," she explains. "Was I touching anything? Was I near any triggers?"

Like many other people with allergies, Mary Ellen became more in tune with her surroundings. That being said, she tells me that her allergies don't factor into how or what she plants for other people. In fact, she tells me that landscapers don't necessarily think about things like pollen load unless their clients ask them to. In her entire career, only one woman told her that she was highly allergic. Mary Ellen was confused about why the woman wanted flower beds at all until she realized that she wanted to stay inside and look at them from afar.

"Most of the time, I am trying to make sure that my clients have seasonal bloom," Mary Ellen says, "and the colors that they want and the type of plants that will stay in scale with the rest of the landscape. That's what I do with design work. I never contemplate, 'Oh, jeez, is this going to make somebody sneeze' or whatever."

I'm curious, at this point in our conversation, if her field at large is concerned at all with respiratory allergies and pollen loads. Do professional landscaping associations or magazines ever touch on the topic?

"I belong to a couple of different landscape associations," Mary Ellen says. "But I've never run into that topic. We go to our conferences to learn about different plants and different tools. But never anything about allergies. It's just occurring to me now that many people that are highly allergic never do any landscaping. And people that inquire about landscaping usually want to remain indoors and allow us to do the work."

She tells me that the current trend in professional landscaping is native plants. In response to environmental concerns, more people are pushing to plant trees, shrubs, and grasses that are native to an area, as opposed to the "exotic" flora that was in vogue for decades. (Remember the Chinese elm from chapter 5?) The native plants produce a lot of pollen, too, Mary Ellen explains, but at least it's usable. The local bees and butterflies and other fauna love it, even if native plants are not any better than exotic plants in terms of pollen load.

"Exotic" plants are anything not naturally occurring in the local ecosystem. As it turns out, there are a few local municipalities that ban certain species based on their production of pollen. Case in point: Pima County, Arizona.

Olive trees, or *Olea europaea,* were first imported to the western coast of the United States in the 1700s by Catholic missionaries. While native to the Mediterranean and some areas in Africa and Asia, the olive tree is well suited to the arid conditions of desert living in Arizona. An olive tree needs little water and is incredibly drought tolerant, perfect for landscaping in drier climates. There's one drawback, however: Olive trees produce a great deal of pollen for two months each year, and many people's immune systems are sensitized to react to it, triggering allergic responses.

Local officials in Pima County noticed that the olive trees were ruining Arizona's reputation as a great place to be for allergy and asthma patients. So in 1984, the county halted any future planting of olive trees (it also, for the record, banned Texas mulberry trees and required property owners to keep Bermuda grass neatly trimmed for the same reasons). It was the first county in the United States to ban a specific species of tree for its pollen load. A year later, Boulder City, Nevada, followed suit. As of this writing, the ban remains in place.

Pima County officials claimed that only three years after the ban went into effect, the air was noticeably clearer. And yet, respiratory allergies in Pima County have not dissipated. Why? There are other types of pollen, like that from mesquite trees, which are native to the region.

When I called Pima County to discuss the ban, no one familiar with it was available to talk to me. During my research for this section, I began to notice something odd. No one who worked in government parks and recreation offices anywhere was willing to discuss pollen with me. And, believe me, I tried. I called several municipalities—

including New York City and Chicago—to no avail. When all is said and done, I suppose pollen is political. There is no "good" answer to my question about what we can do collectively to help seasonal allergy sufferers because there is no getting rid of pollen. In fact, any efforts to curtail it might be seen as somehow not environmentally friendly. So, what to do?

Do we ban non-native trees and grasses? Do we try to regulate how many pollinating trees, grasses, and plants are in any geographic area? Or do we let people with respiratory allergies figure it out for themselves while we focus on getting more dangerous things, like particulate matter, out of the air?

The answers to these questions aren't clear. And at least in the case of Pima County, it doesn't seem as though bans on certain species are all that effective. In the end, it turns out that *nothing* about crafting an environmental allergy policy is simple. But by now, in the final chapter of this book, you probably already knew that it wouldn't be.

THE FUTURE OF ALLERGY

Ultimately, our policies and laws reflect the dominant cultural paradigms of our age. How we think about allergies, how they are depicted in the media we consume, and how much exposure to and education we have about allergic conditions all affect what we, as a society, decide to do about them. Here are the questions that I believe should dominate allergy policy going forward: As global allergy rates continue to rise precipitously over the next few decades, will we work collectively through new laws, regulations, and cultural norms to help prevent or alleviate them? Will we give up some of our habits and traditions to help make the world a more livable space for everyone? Or will we continue to ask individual allergy sufferers to take on the full brunt of responsibility for their conditions themselves? The choices we make will determine how healthy our world is for all of our immune systems for centuries to come.

Irritating Ourselves to Death:
Allergy in the Time of COVID-19

This continued exposure to low levels of toxic agents will eventually result in a great variety of delayed pathological manifestations, creating physiological misery, increasing the medical burden, and lowering the quality of life.[1]

—René Dubos, French American microbiologist, 1966

I struggled with how to end this book on allergy. Over and over again, we've learned how complicated allergies really are. We've learned that allergy is about our vulnerability—both biological and social—and we've learned about the challenges of living in a changing environment. I wanted to leave you on an optimistic note because so much of what you've discovered here is scary and depressing.

But the stark truth is that our overworked immune systems aren't faring well in the twenty-first century. A decrease in overall air quality around the globe—from increased air pollution to higher pollen counts—is slowly making it harder for all of us to breathe. But it's not just climate change and our relationship to our natural environment that may doom us; it's everything about how we live now. Changes in food production and diets, along with increased reliance on antibiotics, are contributing to higher rates of allergy everywhere. New chemical and industrial products are making our skin more irritated.

Everything we've been doing for the last two hundred years (as the new alpha-gal allergy attests) is irritating us—slowly, imperceptibly, constantly—and allergic people are like canaries in the environmental coal mine. They might be the ones suffering more now, but they're a harbinger of things to come for all of us. Allergies are, to quote one allergist, "the model of the health impacts of climate change."

We're literally irritating ourselves to death. The question is: What are we going to do about it? The answer is either (A) Do nothing differently and watch as allergies get worse because our immune systems continue to be both overwhelmed and undertrained for life in the twenty-first century or (B) Realize we're causing the lion's share of the allergy epidemic and start to collectively rethink how we live our day-to-day lives, shifting to more sustainable living and changing our entire relationship to our surroundings.

While I'd like to be more optimistic about the odds of us choosing option B, as any good doctor will tell you, people don't always do the things that are best for them, especially when doing so asks us to radically alter our thinking as well as our behaviors. And yet, if we don't reimagine our relationship to the microscopic worlds around us, where will that lead us?

In January 2020, as the world began to slowly realize we were at the beginning of what would end up being the largest and deadliest global pandemic in more than a century, our relationship to our environment was cast in a new light—particularly in relationship to the invisible things around us. Microscopic particles (both good and bad) are everywhere. Microbes, in particular, have always been our companions; some of them form a critical part of us. It is no exaggeration to argue that we are not wholly human. We are barely even *mostly* human.

Right now, as your eyes scan across this page, you are harboring more microbes than human cells. The entity that is "you" is a collection of microorganisms and cells cooperating to look like and function as a "human being." Remember that description of the Portuguese man-of-war that began our history of allergy? Well, that's a lot like you. You're like a walking collection of bacteria and viruses with a smartphone and shoes. In other words, you're a bazillion cells—human and not—working together symbiotically, just like every other so-called higher organism on the planet.

Dr. David Bass, a leading member of the research teams at the Centre for Environment, Fisheries and Aquaculture Science and the University of Exeter in the United Kingdom, argues that "the vast majority of cells in our bodies are bacterial, not human. Therefore, we are walking ecosystems—interacting communities of many different organisms."[2]

This is neat information to have, you might be thinking, but exactly how does this fact play into our biography of allergy? Well, if the human immune system exists to maintain the balance of helpful and harmful cells throughout the body—the body's natural curator, so to speak—then a person's individual microbiome may be a key not only to better health but also to understanding how the immune system functions in the first place, as Dr. Cathy Nagler and her peers have argued so forcefully throughout this book. If the partial successes of immunotherapy are anything to go by, then studying the microbiome—or how human cells interact with our own internal bacteria and viruses—may help us to unravel the entire mystery of allergies.

If there is a true "cure" for allergies, it must lie in our complicated relationships with and dependence upon what we commonly refer to as "germs." Because it turns out that some germs are our friends, not our enemies. And having the right mixture of microbes—both around us and inside us—is necessary to our health and well-being. We literally cannot live well without them.

As someone who has been studying viruses her whole adult life, this doesn't surprise me. Viruses and bacteria are everywhere; they are the building blocks of life. They exist at the deepest levels of the ocean and in the driest deserts, in environments where nothing else can thrive. So why wouldn't they be necessary to our own health and survival? I find solace in knowing that we're part of the ecosystem and not separate from it. If we can rethink what it means to be human, and learn not just to coexist with microbes, but to foster and nurture our relationships with them, then I think allergies have a decent shot at becoming a thing of the past like smallpox or (at least until recently) polio.

The COVID-19 pandemic has highlighted the stark need for us to better understand how human actions affect the microscopic world, and how our immune systems interact with it. Research during the pandemic suggested that COVID-19 risk increased along with the amount of pollen in the air. The level of pollen could, in fact, explain 44 percent of the variation in COVID-19 infection rates. There were two reasons for this. First, higher pollen levels lead to a weakened immune response. The pollen grains, in effect, allowed the virus particles to evade the already overworked immune system. Imagine pollen and virus as a throng pouring in through the doors of a stadium; it's far easier to pick out and stop virus particles at the gate when they aren't mixed in with pollen grains. Second, virus particles can attach to pollen grains circulating in the air, helping them to float longer and farther than they normally would. These types of complex environmental interactions and their effect on our immune system can make the difference between surviving a pandemic and succumbing to one. Understanding more about our basic immune function and how our immune systems respond to different particles may help us design more effective prevention tools and treatments in the future.

As I write this, the Omicron variant of SARS-CoV-2 is causing COVID-19 cases to tick upward, and the unvaccinated are beginning to fill hospital wards around the globe. Despite this, the world is beginning to emerge from the long quarantine and social distancing that went into effect in March 2020. The vaccines for COVID-19, many of which used groundbreaking mRNA technology, are still effective at preventing serious cases of disease. Immunologists and virologists have learned more about our immune function but worries about the response of our immune systems remain. Researchers wonder what effects social distancing and isolation might have had on the ability of our immune systems to cope with exposures as we reengage with one another. Warnings abound that children might be sicker than usual as they return to school settings, camps, and play-dates because their immune systems are "out of shape" after being quarantined.

The truth is that we don't know what effects this pandemic has had on our bodies. We are all part of an unintended, massive natural experiment. Scientists everywhere are scrambling to keep up with developments. The silver lining to all the death, economic disaster, and social disruption created by SARS-CoV-2 is this: We will come out of it knowing more about our immune systems.

At the start of this pandemic, there were around eight thousand immunologists in the world. My hope is that, following COVID-19 and with the rising rates of allergies around the globe, there will soon be many more. In fact, the immunologists and allergists I met throughout my research for this book are what give me the most hope for the future. Without fail, the experts I spoke with were some of the smartest, most generous, and most dedicated people I have ever had the pleasure of meeting. They are determined to figure out the puzzle of allergic response, to alleviate our immune irritation, and to harness all of our scientific and technological prowess to rebalance and reshape the human relationship to the environment. Knowing that we are in capable and caring hands makes it easier for me to sleep at night. I hope it does the same for you.

Researching and writing this book has changed me significantly. I have begun to look for ways to help my own immune system: I eat more natural foods and fewer processed ones; I get enough sleep and exercise; I stopped taking daily showers and changing my sheets as often; I committed to reducing my carbon footprint; I vote for political candidates who support action on climate change and environmental protections; I use less "stuff" on my skin. I urge you to take the information in this book and rethink your own habits and actions. Together, despite everything we've done to ourselves and the natural world, I still believe that we have time to choose option B.

———

MY FATHER'S DEATH, REVISITED

By now, I understand far better how my father died. I understand, too, all the ways that I am connected to him, both genetically and via our personalities. My dad was an irritable guy—two tours in Vietnam will do that to a person. He was often anxious and depressed, which meant that he ate too much, smoked too much, and drank too much. In other words, he dealt with life in the twentieth century the way most of us deal with life in the twenty-first.

A bee sting killed my father, but that's not the only thing that killed him. If he hadn't been a smoker, he wouldn't have had his window open that day, and the bee wouldn't have flown in. He didn't carry an EpiPen because it was too expensive. He was a smoker because he was stressed out, trying to make ends meet. He was trying to make ends meet because he didn't have a college education, because he enlisted in the army at eighteen instead. And his reasons for doing that are his own.

Now that I am older than he was when he died, I know how complicated life can be. I am often stressed out, too, and I do all kinds of things that don't make sense—like not carrying an EpiPen myself (though I'll probably remedy that soon). Allergy fascinates me because it's an illness that you develop just from being born and living your life in this out-of-balance world. Allergy is a strange "illness." It's not caused by anything you've done, but at the same time, it's everyone's fault. You're not sick, but you're not well either. If your immune response gets triggered by the wrong thing, it will kill you while trying to protect you.

I think my dad understood all this on an intuitive level, having lived through a disastrous war and growing up in the tumultuous 1960s. He understood miscommunication and what happens when you fight the wrong battles. I began this journey because I wanted to understand what happened to my dad, and what was happening to me and many of my friends. In the beginning, I only wanted to diagnose the problem of allergy in America. But at the end, I think what I've started to glimpse is really the story of what is happening to all of us—to humanity itself—as we grapple with how we've altered our environment and continue to reshape the worlds around us. Allergy is ultimately about our human vulnerability, both biological and social. For better or for worse, allergies prove that we're all in this increasingly irritated world together. And it will take all of us working together to effectively treat our condition.

ACKNOWLEDGMENTS

Writing a book is often a years-long group effort. In my case, I have spent more than five years researching and writing. I first came up with the idea for this book when I was complaining to my good friend and fellow medical anthropologist and writer Eric Plemons, who finally reminded me that I was a scholar and if there weren't any good books on the problem of allergies, then I could write one myself . . . and so I did. Eric's endless patience and quality advice have helped make this book what it is today. My good friend and colleague and fellow writer Billy Middleton eyeballed different drafts of this book, which only helped me to make it stronger. I also want to thank the former students who helped me forage for information and conduct interviews in the earliest stages of this book, especially Olivia Schreiber (who is on her way to becoming an MD right now and I couldn't be prouder).

My indefatigable agent, Isabelle Bleecker, read everything I wrote and also took every one of my panicked phone calls . . . even after five on Fridays. Everyone should be as lucky to have a literary agent like her. My wonderful editor, Caitlin McKenna, understood the scope and ambition of this project from its very inception and helped me to shape it into the book we both knew it could be. Everyone should be so lucky to have an editor as generous as she is.

My extended team at Random House is equally wonderful. Goodness knows how Noa Shapiro does all that she does, but she does it with aplomb. Thanks also to my lovely and hardworking marketing team getting the word out: Ayelet Durantt, Monica Stanton, and

Windy Dorrestyn. And much gratitude to the design team that created this beautiful physical object that you're holding: Simon Sullivan and Greg Mollica. And of course, many thanks to the production team itself: Rebecca Berlant, Richard Elman, and Ada Yonenaka. Thanks to my publicity team for making sure people hear more about the problem of allergies: London King, Maria Braeckel, and Greg Kubie. And last, but not least, thanks to everyone who supported this book at Random House: publisher Andy Ward, deputy publisher Tom Perry, assistant publishing director Erica Gonzalez, and nonfiction editorial director Ben Greenberg. I feel extremely grateful to be part of the Random House team.

At the start of this project, I was fortunate to receive a National Endowment for the Humanities Public Scholars Award. (Any views, findings, conclusions, or recommendations expressed in this book do not necessarily reflect those of the National Endowment for the Humanities.) It allowed me to take a much-needed year off from teaching in order to research the lion's share of this book. It also helped to cover the cost of a survey on American beliefs and attitudes about allergies. And if you don't already know this, surveys are insanely expensive, so I felt lucky to be able to conduct one of my own with the NEH's generous funding. In that vein, let me also thank my friend Will Hart and PSB Insights for letting me throw some questions about allergy onto their pulse survey—for free. It's a generosity that I will never forget. I also want to thank the amazing librarians and staff at both the Drs. Barry and Bobbi Coller Rare Book Reading Room at the New York Academy of Medicine and the National Library of Medicine. Y'all helped me locate some of the rarest and most critical texts on the early history of allergy. Librarians are, it has to be said, the unsung heroes of scholarly research. (In that vein, please support your libraries! They always need more funding.)

I have the utmost gratitude for every scientist, clinician, and patient who let me pick their brains about allergy. The experts I spoke to over the last five years are some of the kindest, most generous people I've ever had the pleasure to interview. The patients were incredibly open and willing to share their experiences with me and made this a more personal and effective book. I can't thank all of them enough. I would, however, like to single out two of the scientists who went above and beyond. First, Steve Galli, who patiently read the entire manuscript and carefully and kindly corrected any scientific errors I had made. Second, Cathy Nagler, who read the entire manuscript *twice* to make sure I was explaining all of the science correctly—which in our business is akin to sainthood. See what I mean about allergy experts being the best?

Special thanks to three people who shaped me as a writer and thinker. First, Jane Harrigan, the former chair of journalism at UNH, who somehow never gave up on the idea of me making it as a writer and still finds time to occasionally send me encouraging messages. I guess she was right. Second, Stefan Helmreich, anthropologist extraordinaire, who convinced me that if I got a PhD in anthropology, I could study whatever I wanted and have fun doing it. He was right. And third, Vincanne Adams, a terrific medical anthropologist and one of my thesis advisors at UC-Berkeley/UCSF, who counseled me to embrace my journalistic side and use it to become a better and more visible scholar of medicine. She was also spot on. These professors are living proof that teachers can alter the course of a person's life in myriad ways. I wouldn't be writing this now if I hadn't had such amazing role models.

And finally, to my colleagues, my friends, and my partner, Max—all the thanks in the world. No one should have to put up with a writer trying to write an impossible book and harping on about it incessantly . . . but all y'all did. I'd like to say that you won't ever have to go through any of this stuff and nonsense with me again, but we all know that you will.

NOTES

Prologue · Everything That Irritates Us

1. David B. K. Golden, "Insect Allergy," in *Middleton's Allergy Essentials,* ed. Robyn E. O'Hehir, Stephen T. Holgate, and Aziz Sheikh (Amsterdam: Elsevier, 2017), 377.
2. Centers for Disease Control, "QuickStats: Number of Deaths from Hornet, Wasp, and Bee Stings, Among Males and Females—National Vital Statistics System, United States, 2000–2017," *Morbidity and Mortality Weekly Report* 68, no. 29 (July 26, 2019): 649.

Chapter 1 · What Allergy Is (and Isn't)

1. Ruby Pawankar, Giorgio Walkter Canonica, Stephen T. Holgate, Richard F. Lockey, "White Book on Allergy 2011–2012 Executive Summary," *World Allergy Organization.* https://www.worldallergy.org/UserFiles/file/WAO-White-Book-on-Allergy_web.pdf.
2. The names of most of the allergy patients profiled in this book have been changed to protect their privacy. For the few exceptions to this rule, both first and last names are used in the text.
3. We'll examine this history much more deeply in chapter 4 on genetics, inheritance, and allergy as a "normal" immune reaction.
4. J. M. Igea, "The History of the Idea of Allergy," *Allergy* 68, no. 8 (August 2013): 966–73.
5. Warwick Anderson and Ian R. Mackay, *Intolerant Bodies: A Short History of Autoimmunity* (Baltimore: Johns Hopkins University Press, 2014), 28.
6. Antibodies were visible under microscopes and scientists understood that they played a key role in fighting off bacteria. However, the term

"antibody" in the early 1900s was significantly different from our use of the term today.

7. As the term "allergy" grew more popular, Pirquet grew increasingly frustrated with the conflation of his original term with "hypersensitiv-ity" or "overreaction." Pirquet felt it was a mistake to see allergy as *just* a hypersensitive immune system response since it altered his fundamen-tal theory of allergy itself. Tired of repeated attempts to correct his fel-low scientists' usage, Pirquet eventually abandoned the term altogether. The meaning of "allergy" would never again refer to positive biological reactions like immunity.

8. The first scientific journal dedicated to allergy was the *Journal of Al-lergy*, published in 1929. It is still one of the field's leading publications and is titled *The Journal of Allergy and Clinical Immunology*.

9. Warren T. Vaughan, *Allergy and Applied Immunology: A Handbook for Physician and Patient, on Asthma, Hay Fever, Urticaria, Eczema, Migraine and Kindred Manifestations of Allergy* (St. Louis: C. V. Mosby, 1931), 43.

10. George W. Bray, *Recent Advances in Allergy (Asthma, Hay-Fever, Ec-zema, Migraine, Etc.)* (Philadelphia: P. Blakiston's, 1931), 5.

11. William Sturgis Thomas, "Notes on Allergy, circa 1920–1939." Two binders of private notes available in the Drs. Barry and Bobbi Coller Rare Book Reading Room at the New York Academy of Medicine. Many thanks to the diligence of the rare books librarian for her help in locating them and bringing them to my attention.

12. In fact, during the nineteenth century, hay fever was first considered to be another infectious disease, similar to the common cold, but no one could replicate Koch's postulates (the microbe must be found only in diseased individuals, has to be cultivated from samples taken from dis-eased individuals, and then those cultures should be capable of causing disease in a healthy individual) on an allergen, thereby scientifically proving that a living microbe caused the affliction.

13. G. H. Oriel, *Allergy* (London: Bale & Danielsson, 1932), 5.

14. Igea, "History of the Idea of Allergy."

15. Arthur F. Coca, *Asthma and Hay Fever in Theory and Practice. Part I: Hypersensitiveness, Anaphylaxis, Allergy* (Springfield, Ill.: C. C. Thomas, 1931), 4.

16. Thomas A. E. Platts-Mills, Peter W. Heymann, Scott P. Commins, and Judith A. Woodfolk, "The Discovery of IgE 50 Years Later," *Annals of Allergy, Asthma & Immunology* 116, no. 3 (2016): 179–82.

Chapter 2 · How Allergy Diagnosis Works (or Doesn't)

1. Occasionally, someone will have what is known as a "late-phase reaction" to a skin-prick or intradermal test—a reaction that begins one or two hours after the skin tests are performed, peaking at six to twelve hours post–skin test. These late-phase reactions are often not recorded at all, and their underlying biological mechanisms and significance are not well understood.

2. Anca Mirela Chiriac, Jean Bousquet, and Pascal Demoly, "Principles of Allergy Diagnosis," in *Middleton's Allergy Essentials,* ed. Robyn E. O'Hehir, Stephen T. Holgate, and Aziz Sheikh (Amsterdam: Elsevier, 2017), 123.

3. Samuel M. Feinberg, *Asthma, Hay Fever and Related Disorders: A Guide for Patients* (Philadelphia: Lea & Febiger, 1933), 48.

4. Warren T. Vaughan, *Allergy and Applied Immunology: A Handbook for Physician and Patient, on Asthma, Hay Fever, Urticaria, Eczema, Migraine and Kindred Manifestations of Allergy* (St. Louis: C. V. Mosby, 1931).

5. The downside of the P-K serum test is that it can transfer other blood-borne diseases to the nonallergic test subject (such as hepatitis or AIDS). This is partially why its usage was limited and strictly controlled.

6. William Sturgis Thomas, "Notes on Allergy, circa 1920–1939." Two binders of private notes available in the Drs. Barry and Bobbi Coller Rare Book Reading Room at the New York Academy of Medicine.

7. Feinberg, *Asthma, Hay Fever and Related Disorders.*

8. Arthur F. Coca, *Asthma and Hay Fever in Theory and Practice. Part I: Hypersensitiveness, Anaphylaxis, Allergy* (Springfield, Ill.: C. C. Thomas, 1931), 322–29.

9. Albert Rowe, *Food Allergy: Its Manifestations, Diagnosis and Treatment, with a General Discussion of Bronchial Asthma* (Philadelphia: Lea & Febiger, 1931), 21.

10. Guy Laroche, Charles Richet, fils, and François Saint-Girons, *Alimentary Anaphylaxis (Gastro-intestinal Food Allergy)* (Berkeley: University of California Press, 1930).

11. Rowe, *Food Allergy,* 20.

12. Chiriac, Bousquet, and Demoly, "Principles of Allergy Diagnosis," 120.

13. T. Ruethers, A. C. Taki, R. Nugraha, et al., "Variability of allergens in commercial fish extracts for skin prick testing" in *Allergy* 2019 (74): 1352–1363.

14. Mahboobeh Mahdavinia, Sherlyana Surja, and Anju T. Peters, "Prin-

ciples of Evaluation and Treatment," in *Patterson's Allergic Diseases*, 8th ed., ed. Leslie C. Grammer and Paul A. Greenberger (Philadelphia: Wolters Kluwer, 2018), 160–62.

15. Mahdavinia, Surja, and Peters, "Principles of Evaluation and Treatment," 159.

16. Interestingly, Dr. Lio told me that adult-onset atopic dermatitis was thought to be a myth when he was training. It's more accepted today. Contact skin allergy, on the other hand, has always been considered to be adult-onset because of occupational contact skin allergies, wherein people develop a sensitivity to a substance through repeated contact (like latex allergy in healthcare workers).

17. Adnan Custovic, "Epidemiology of Allergic Diseases," in *Middleton's Allergy Essentials*, ed. Robyn E. O'Hehir, Stephen T. Holgate, and Aziz Sheikh (Amsterdam: Elsevier, 2017), 54: "Most epidemiologic studies define atopic sensitization as a positive allergen-specific serum IgE . . . or a positive skin-prick test. . . . However, positive 'allergy' tests indicate only the presence of allergen-specific IgE (either in serum or bound to the membrane of mast cells in the skin), and are not necessarily related to the development of the clinical symptoms upon allergen exposure. Indeed, a sizeable proportion of individuals with positive allergy tests have no evidence of allergic disease." In other words, you can have a clinically observable sensitivity without ever experiencing a reaction. Some studies have suggested that the size of the wheal formed by the skin-prick test, plus the presence of IgE antibodies, is more predictive of symptoms or "allergic disease."

18. Scott H. Sicherer and Hugh A. Sampson, "Food Allergy: Epidemiology, Pathogenesis, Diagnosis, and Treatment," *Journal of Allergy and Clinical Immunology* 133, no. 2 (February 2014): 295.

19. Sicherer and Sampson, "Food Allergy," 296. Even if an OFC is not performed, Sicherer and Sampson argue that standard skin-prick tests and sIgE tests can "go a long way in assisting in diagnosis."

20. An expert panel at the National Institute of Allergy and Infectious Diseases "identified 4 categories of immune-mediated adverse food reactions (eg, food allergies), namely IgE-mediated, non-IgE mediated, mixed, or cell-mediated reactions. . . . There are a number of disorders that are not food allergies but might appear similar" (Sicherer and Sampson, "Food Allergy," 294). For example, celiac disease is non-IgE mediated, but skin contact allergy is cell mediated.

21. Chiriac, Bousquet, and Demoly, "Principles of Allergy Diagnosis," 123.

22. Chiriac, Bousquet, and Demoly, "Principles of Allergy Diagnosis," 123.

23. As a matter of interest, a positive skin test is necessary to start immunotherapy; however, skin tests cannot be used to assess the success of immunotherapy or to determine when to halt treatment because they show only sensitivity, not the presence of an allergy. If immunotherapy works, a patient will no longer have symptoms—or the allergy—but will retain the sensitivity, or the propensity toward an allergy. Hence, immunotherapy never changes the results of a skin test.

Chapter 3 · Our Allergic World: Measuring the Rise of Allergic Disease

1. Adnan Custovic, "Epidemiology of Allergic Diseases," in *Middleton's Allergy Essentials,* ed. Robyn E. O'Hehir, Stephen T. Holgate, and Aziz Sheikh (Amsterdam: Elsevier, 2017), 52.
2. Custovic, "Epidemiology of Allergic Diseases."
3. Custovic, "Epidemiology of Allergic Diseases."
4. Custovic, "Epidemiology of Allergic Diseases."
5. Custovic, "Epidemiology of Allergic Diseases."
6. Custovic, "Epidemiology of Allergic Diseases."
7. Lymari Morales, "More Than 10% of U.S. Adults Sick with Allergies on a Given Day," Gallup News, November 17, 2010, https://news.gallup.com/poll/144662/adults-sick-allergies-given-day.aspx.
8. R. S. Gupta et al., "Prevalence and Severity of Food Allergies Among US Adults," *JAMA Network Open* 2, no. 1 (2019): e185630.
9. Scott H. Sicherer and Hugh A. Sampson, "Food Allergy: Epidemiology, Pathogenesis, Diagnosis, and Treatment," *Journal of Allergy and Clinical Immunology* 133, no. 2 (February 2014): 291–302.
10. Custovic, "Epidemiology of Allergic Diseases," 61.
11. Custovic, "Epidemiology of Allergic Diseases," 62.
12. Severity is incredibly difficult to measure because it relies on patients' subjective experiences of their allergies. There are currently no adequate measures of allergy severity other than self-reporting and clinical observations. That said, most of the allergy patients I interviewed suggested to me that as the years unfolded their allergies were getting worse in terms of severity. This was especially true of seasonal allergy patients.
13. A. B. Conrado et al., "Food Anaphylaxis in the United Kingdom: Analysis of National Data, 1998–2018," *The BMJ* 372 (2021): n251.
14. Custovic, "Epidemiology of Allergic Diseases," 61–62.

Chapter 4 · Allergic Inheritance: Allergies as a "Normal" Immune Response

1. The bulk of my retelling of Portier and Richet's discovery here is reliant on two sources: Charles D. May, "The ancestry of allergy: being an account of the original experimental induction of hypersensitivity recognizing the contribution of Paul Portier" in *Journal of Allergy and Clinical Immunology* 75, no. 4 (April 1985): 485–495. And Sheldon G. Cohen and Myrna Zeleya-Quesada, "Portier, Richet, and the discovery of anaphylaxis: A centennial," in *The Allergy Archives: Pioneers and Milestones,* Volume 110, Issue 2 (2002): 331–336.

2. In addition to his own research, Portier would become the director of the Institut océanographique (funded by Prince Albert I) when it opened in 1908, eventually supervising more than one hundred dissertations in marine biology.

3. Unfortunately, Richet also believed in the biological inferiority of non-white races. He maintained an avid interest in eugenics until his death in 1935.

4. Humphry Rolleston, *Idiosyncrasies* (London: Kegan, Paul, Trench, Trubner & Co., 1927).

5. Laurence Farmer and George Hexter, *What's Your Allergy?* (New York: Random House, 1939), 8–9. Quoted later in the same text, Hutchinson laments, "Where idiosyncrasies were concerned, medicine was playing blindman's bluff" (17).

6. Interestingly, Arthur Coca questioned the available "statistical studies of the hereditary nature" of atopy. First, the people being asked about their family medical history "must be essentially intelligent and sufficiently acquainted with the meaning of the terms" to answer the questions. After all, if one didn't understand what hay fever or asthma was, how would they be able to assess if their relatives had the condition? Second, only people who actually interacted with relatives could be expected to know the answers to these questions. For example, it could not be known for certain whether a long-dead ancestor had asthma. Third, if the patient was too young, then all of their symptoms might not have had time to manifest, which would throw the statistics off. Also, if the person being interviewed was living in the United States but originally came from Europe, one might not get relevant data at all about sensitivity to things that existed only in America or only in Europe. Along the same line of thinking, if someone had never been exposed to the allergen, then one wouldn't know if they were sensitive to it. It seems that the difficulty in obtaining good stats from self-reported

surveys has plagued allergy researchers from the very beginnings of the field of allergology and is not a new problem at all. Arthur F. Coca, *Asthma and Hay Fever in Theory and Practice. Part I: Hypersensitiveness, Anaphylaxis, Allergy* (Springfield, Ill.: C. C. Thomas, 1931): 42.

7. William Sturgis Thomas, "Notes on Allergy, circa 1920–1939." Two binders of private notes available in the Drs. Barry and Bobbi Coller Rare Book Reading Room at the New York Academy of Medicine.

8. Guy Laroche, Charles Richet, fils, and François Saint-Girons, *Alimentary Anaphylaxis (Gastro-intestinal Food Allergy)* (Berkeley: University of California Press, 1930).

9. Rolleston, *Idiosyncrasies,* 42.

10. W. Langdon-Brown, "Allergy, Or, Why One Man's Meat Is Another's Poison," Abstract of Lecture Given Before the Cambridge University Medical Society, October 19, 1932.

11. Fun fact that's not so fun: Children's maladies were often thought to be caused by their mothers' genetics or behavior. For instance, children with severe asthma were often separated from their mothers because it was thought that the mothers' anxiety or neuroses were contributing to their children's attacks. Bias against mothers and women within medicine was rife during this period. Unfortunately, gender bias in medical diagnoses still contributes to differential outcomes for female patients. For a good overview of these issues, see Maya Dusenbery, *Doing Harm: The Truth About How Bad Medicine and Lazy Science Leave Women Dismissed, Misdiagnosed, and Sick* (New York: HarperOne, 2018).

12. Largely because anaphylaxis was only studied in animals in laboratories and not in humans. Controlled experiments in a lab can seem more standardized than observations in the real world.

13. Arthur F. Coca, *Asthma and Hay Fever in Theory and Practice. Part I: Hypersensitiveness, Anaphylaxis, Allergy* (Springfield, Ill.: C. C. Thomas, 1931).

14. Walter C. Alvarez, *How to Live with Your Allergy* (New York: Wilcox & Follett, 1951).

15. Samuel M. Feinberg, *Allergy Is Everybody's Business* (Chicago: Blue Cross Commission, 1953).

16. Asthma is more prevalent in boys but more prevalent and severe in adult females. Testosterone suppresses the production of a type of immune cell that triggers asthma (ILC2). Estrogen is inflammatory, which is why women often report changes during pregnancy.

17. H. Milgrom and H. Huang, "Allergic Disorders at a Venerable Age: A Mini-review," *Gerontology* 60, no. 2 (2014): 99–107. Aging immune systems and changes in bacterial makeup can lead to worsening aller-

gies in the elderly; 5–10 percent of the elderly have allergic disease and rates are rising.

18. F. Hörnig et al., "The LINA Study: Higher Sensitivity of Infant Compared to Maternal Eosinophil/Basophil Progenitors to Indoor Chemical Exposures," *Journal of Environmental and Public Health* (2016). Higher concentrations of plasticizers equate to a greater risk of developing allergies (as measured by benzyl butyl phthalate, or BBP, in the urine of mothers). Exposure to phthalates during pregnancy and breastfeeding cause epigenetic changes to repressors for Th2.

　　For a discussion of sensitivity transfers, see Rasha Msallam et al., "Fetal Mast Cells Mediate Postnatal Allergic Responses Dependent on Maternal IgE," *Science* 370 (November 20, 2020): 941–50. Mothers (at least in mice models) can pass allergies to offspring. If they are exposed to an allergen (in this case, ragweed) while pregnant, IgE antibodies can travel through the placenta to the fetus and bind with fetal mast cells. Once born, those offspring are more prone to an allergen reaction on the first exposure to ragweed (as opposed to another allergen, in this case, dust mites). The sensitivity transfers lasted only a few weeks, and most had disappeared by six weeks. But this research (done in Singapore by scientists at A*STAR and Duke-NUS Medical School) shows that sensitivity could theoretically be transferred in humans in much the same manner.

19. Åsa Johansson, Mathias Rask-Andersen, Torgny Karlsson, and Weronica E. Ek, "Genome-Wide Association Analysis of 350000 Caucasians from the UK Biobank Identifies Novel Loci for Asthma, Hay Fever and Eczema," *Human Molecular Genetics* 28, no. 23 (2019): 4022–41. A full forty-one of those gene segments had not yet been identified in other studies. The study was conducted using the United Kingdom's Biobank and 23andMe by Uppsala University and SciLifeLab in Sweden.

20. The link between the filaggrin variation and allergic conditions had been posited before, but this was the first study of a birth cohort.

21. Hans Bisgaard, Angela Simpson, Colin N. A. Palmer, Klaus Bønnelykke, Irwin Mclean, Somnath Mukhopadhyay, Christian B. Pipper, Liselotte B. Halkjaer, Brian Lipworth, Jenny Hankinson, Ashley Woodcock, and Adnan Custovic. "Gene-Environment Interaction in the Onset of Eczema in Infancy: Filaggrin Loss-of-Function Mutations Enhanced by Neonatal Cat Exposure" in PLoS Med. 2008 Jun; 5(6): e131.

22. Unfortunately, this type of fix wouldn't necessarily mean that we can prevent eczema from developing at all. Although most adults with eczema had the condition as children, about 25 percent of those with eczema report first experiencing their initial symptoms in adulthood,

according to the National Eczema Association. This is typically referred to as "adult-onset eczema."

23. Interestingly, environmental allergens like those from house dust mites and cockroaches may enter the body with greater ease in a proportion of young children who are vulnerable because of their leaky skin, triggering eczema and asthma. Dr. Mukhopadhyay's study with cats could stimulate further birth cohort studies that help tease out these links between specific environmental exposures and leaky skin.

24. Other NIH scientists have also found that a gene called *BACH2* may play a role in the development of both allergic and autoimmune disorders by regulating the immune system's reactivity. A genome-wide association study that analyzed samples from autoimmune patients first flagged the gene as a possible regulator of inflammatory immune responses. In a 2013 study, NIH researchers found that the expression of *BACH2* was a key factor in how the immune system T cells responded to antigens—either by becoming inflamed or by regulating the response. In an NIH press release for the study, Nicholas P. Restifo, the principal investigator on the project, explained: "It's apt that the gene shares its name with the famous composer Bach, since it orchestrates many components of the immune response, which, like the diverse instruments of an orchestra, must act in unison to achieve symphonic harmony." The study referenced can be found here: R. Roychoudhuri et al., "Bach2 Represses Effector Programmes to Stabilize Treg-mediated Immune Homeostasis," *Nature*, online, June 2, 2013.

25. S. H. Sicherer, T. J. Furlong, H. H. Maes, R. J. Desnick, H. A. Sampson, B. D. Gelb, "Genetics of peanut allergy: a twin study," *Journal of Allergy and Clinical Immunology* 106 (July 2000) (1 Pt 1): 53–56.

26. The study was performed by Jonathan Kipnis; Dr. Milner summarized it for me during our interview at the NIH campus in 2019. J. Herz, Z. Fu, K. Kim, et. al., "GABAergic neuronal IL-4R mediates T cell effect on memory," in *Neuron* 109, no. 22 (November 17, 2021): 3609–3618.

27. A. A. Tu, T. M. Gierahn, B. Monian, et al., "TCR sequencing paired with massively parallel 3' RNA-seq reveals clonotypic T cell signatures," in *Nature Immunology* 20 (2019): 1692–1699.

28. G. William Wong et al., "Ancient Origin of Mast Cells," *Biochemical and Biophysical Research Communications* 451, no. 2 (2014): 314–18.

29. Hadar Reichman et al., "Activated Eosinophils Exert Antitumorigenic Activities in Colorectal Cancer," *Cancer Immunology Research* 7, no. 3 (2019): 388–400. This Tel Aviv University study found that eosinophils may be helpful in the fight against colon cancer by eliminating malig-

nant cells. In 275 patient tumor samples, the higher the number of eosinophils, the less severe the cancer.

30. Martin Metz et al., "Mast Cells Can Enhance Resistance to Snake and Honeybee Venoms," *Science* 313, no. 5786 (2006): 526–30.

31. For those of you who are wondering about Richet's early experiments with the Portuguese man-of-war toxins, not all toxins are created alike, chemically speaking. Because Richet didn't have access to modern scientific technology, he probably would not have been able to measure any minimal protective effect that mast cell activation may have provided, and in any case, the complicated function of mast cells hadn't even been worked out yet.

32. Galli suggests that this is why no one has even tried to come up with an antibody-based treatment for venom—they couldn't make money from it.

Chapter 5 · Nature Out of Whack

1. Charles H. Blackley, *Experimental Researches on the Causes and Nature of Catarrhus Aestivus (Hay-Fever or Hay-Asthma)* (London: Baillière, Tindall & Cox, 1873). All of the discussion of Blackley that follows is culled from this book—the original publication of the entirety of his research on pollen and hay fever.

2. Laurence Farmer and George Hexter, *What's Your Allergy?* (New York: Random House, 1939). For a long time, it was considered impossible for such a nondescript substance as house dust to be able to cause allergies. Dr. Robert Cooke, quoted in this Farmer and Hexter text and writing in 1917, proposed a detailed case to prove the connection, but it took years for other practitioners to accept that an environmental substance could cause a reaction.

3. He also thought that certain types of pollen might be able to pass through mucous membranes into circulation in the body, causing other symptoms.

4. Blackley also performed experiments with a kite that showed that pollen was present in much greater amounts in the higher strata of the air than nearer to the ground. This led him to believe that pollen could be carried vast distances, causing the disease of hay fever at places far removed from hayfields, meadows, or other vegetation. Still, the mountain air was likely clear of pollen due to the lack of vegetation—and differing types—at higher altitudes. But Blackley couldn't prove this since he couldn't abscond to the mountains to run his experiments while he was also busy running his medical practice.

5. Blackley repeated his experiment on the outskirts of the city and then a bit more into the center of the city. Although the pollen counts were not as high as in the meadows, still they rose and fell in a similar pattern and produced similar symptoms.

6. August A. Thommen, *Asthma and Hay Fever in Theory and Practice. Part III: Hay Fever* (Springfield, Ill.: C. C. Thomas, 1931).

7. Farmer and Hexter, *What's Your Allergy?* Germ theory for allergy was first proposed by a physicist, Hermann von Helmholtz, who was himself a hay fever sufferer and tested his own sputum, finding germs.

8. Anna has since retired and is enjoying a life devoid of staring into microscropes for hours on end.

9. Air quality monitoring and research actually began in the 1940s but got a real boost after Congress passed the Clean Air Act of 1970.

10. Denise J. Wooding et al., "Particle Depletion Does Not Remediate Acute Effects of Traffic-Related Air Pollution and Allergen: A Randomized, Double-Blind Crossover Study," *American Journal of Respiratory and Critical Care Medicine* 200, no. 5 (2019): 565–74.

11. Mark Jackson, *Allergy: The History of a Modern Malady* (London: Reaktion Books, 2009). Mark Jackson's book tracks this historical transition. Here in the United States, children born into poverty are at a much greater risk of developing asthma, largely due to their environmental risk factors. We'll look at the association between socioeconomic status and allergy risk in more detail in chapter 6.

12. World Health Organization, "Asthma Fact Sheet," May 11, 2022, https://www.who.int/news-room/fact-sheets/detail/asthma.

13. While researching this book, I had a conversation with an Uber driver in New Orleans about how allergies and asthma had gotten worse after Hurricane Katrina. The culprit, he said, was all the mold.

14. A. Sapkota et al., "Association Between Changes in Timing of Spring Onset and Asthma Hospitalization in Maryland," *JAMA Network Open* 3, no. 7 (2020).

15. S. C. Anenberg, K. R. Weinberger, H. Roman, J. E. Neumann, A. Crimmins, N. Fann, J. Martinich, P. L. Kinney, "Impacts of oak pollen on allergic asthma in the United States and potential influence of future climate change," in *Geohealth* 1, no. 3 (May 2017): 80–92.

16. Nathan A. Zaidman, Kelly E. O'Grady, Nandadevi Patil, Francesca Milavetz, Peter J. Maniak, Hirohito Kita, Scott M. O'Grady, "Airway epithelial anion secretion and barrier function following exposure to fungal aeroallergens: Role of oxidative stress," in *American Journal of Physiology–Cell Physiology* 313 (2017): C68–C79.

Chapter 6 · Are We Doing This to Ourselves?
The Modern Lifestyle and Allergy

1. George W. Bray, Recent Advances in Allergy, 1931: 46.
2. William Sturgis Thomas, "Notes on Allergy, circa 1920–1939." Two binders of private notes available in the Drs. Barry and Bobbi Coller Rare Book Reading Room at the New York Academy of Medicine.
3. Warren T. Vaughan, *Allergy and Applied Immunology: A Handbook for Physician and Patient, on Asthma, Hay Fever, Urticaria, Eczema, Migraine and Kindred Manifestations of Allergy* (St. Louis: C. V. Mosby, 1931).
4. Samuel M. Feinberg, *Allergy in General Practice* (London: Henry Kimpton, 1934): 32.
5. Laurence Farmer and George Hexter, *What's Your Allergy?* (New York: Random House, 1939): 182. Interesting aside: Of all of the case studies Farmer and Hexter used to prove their point, only one was a man.
6. Arthur Coca, *Asthma and Hay Fever in Theory and Practice* (Springfield, Ill.: C. C. Thomas, 1931): 214–218.
7. Albert Rowe, *Food Allergy: Its Manifestations, Diagnosis and Treatment, with a General Discussion of Bronchial Asthma* (Philadelphia: Lea & Febiger, 1931), 21.
8. There's a much longer, and more involved, history of gender and racial bias within medicine that tracks with early conceptualizations of allergy as a disease of the "weak." The history of medicine is rife with examples—hysteria and chronic fatigue syndrome being just two that readers may be more familiar with—and while I don't have the space to do it justice here, a Google Scholar search for medical bias will result in thousands of scholarly articles on the topic.
9. Walter C. Alvarez, *How to live with your allergy* (Mayo Foundation, 1951): 36.
10. Samuel Feinberg, *One Man's Food* (Chicago: Blue Cross Commission, 1953): 2–3.
11. Allergy Foundation of America, *Allergy: its mysterious causes and modern treatment* (1967). This pamphlet is available at the New York Academy of Medicine.
12. Robert Cooke, *Allergy in Theory and Practice* (Philadelphia: Saunders, 1947): 323.
13. Michigan State University, "Here's How Stress May Be Making You Sick," ScienceDaily, January 10, 2018, www.sciencedaily.com /releases/2018/01/180110132958.htm; Helene Eutamene, Vassilia Theodoru, Jean Fioramonti, and Lionel Bueno, "Acute Stress Modulates

the Histamine Content of Mast Cells in the Gastrointestinal Tract Through Interleukin-1 and Corticotropin-Releasing Factor Release in Rats," *Journal of Physiology* 553, pt. 3 (2003): 959–66, doi:10.1113 /jphysiol.2003.052274; Mika Yamanaka-Takaichi et al., "Stress and Nasal Allergy: Corticotropin-Releasing Hormone Stimulates Mast Cell Degranulation and Proliferation in Human Nasal Mucosa," *International Journal of Molecular Sciences* 22, no. 5 (2021): 2773, doi: 10.3390/ijms22052773.

14. K. Harter et al., "Different Psychosocial Factors Are Associated with Seasonal and Perennial Allergies in Adults: Cross-Sectional Results of the KORA FF4 Study," *International Archives of Allergy and Immunology* 179, no. 4 (2019): 262–72. The average age of study participants was sixty-one. It would be interesting to see what the correlations would be in different age groups, or by gender.

15. It's a vicious cycle, and one that patients aren't always able to break by themselves. We'll revisit the social aspect of allergies in chapter 10.

16. D. P. Strachan, "Hay Fever, Hygiene, and Household Size," *BMJ* 299 (1989): 1259–60.

17. Onyinye I. Iweala and Cathryn R. Nagler, "The Microbiome and Food Allergy," *Annual Review of Immunology* 37 (2019): 379.

18. G.A.W. Rook, C. A. Lowry, and C. L. Raison, "Microbial 'Old Friends,' Immunoregulation and Stress Resilience," *Evolution, Medicine, and Public Health* 1 (January 2013): 46–64.

19. Erika von Mutius, "Asthma and Allergies in Rural Areas of Europe," *Proceedings of the American Thoracic Society* 4, no. 3 (2007): 212–16: "These findings suggest that dust from stables of animal farms contains strong immune-modulating substances and that these as yet unknown substances suppress allergic sensitization, airway inflammation, and airway hyperresponsiveness in a murine model of allergic asthma."

20. J. Riedler et al., "Exposure to Farming in Early Life and Development of Asthma and Allergy: A Cross-Sectional Survey," *Lancet* 358, no. 9288 (October 6, 2001): 1129–33. Their findings were as follows: Exposure to stables and consumption of farm milk before the age of one was associated with lower frequencies of asthma (1 percent versus 11 percent), hay fever (3 percent versus 13 percent), and atopic sensitization (12 percent versus 29 percent). The lowest levels of asthma were associated with continual exposure to stables until age five.

21. Christophe P. Frossard et al., "The Farming Environment Protects Mice from Allergen-Induced Skin Contact Hypersensitivity," *Clinical & Experimental Allergy* 47, no. 6 (2017): 805–14.

22. Hein M. Tun et al., "Exposure to Household Furry Pets Influences the

Gut Microbiota of Infant at 3–4 Months Following Various Birth Scenarios," *Microbiome* 5, no. 1 (2017).

23. G. T. O'Connor et al., "Early-Life Home Environment and Risk of Asthma Among Inner-City Children," *Journal of Allergy and Clinical Immunology* 141, no. 4 (2018): 1468–75.

24. J.K.Y. Hooi et al., "Global Prevalence of *Helicobacter pylori* Infection: Systematic Review and Meta-Analysis," *Gastroenterology* 153, no. 2 (August 2017): 420–29.

25. M. J. Blaser, Y. Chen, and J. Reibman, "Does *Helicobacter pylori* Protect Against Asthma and Allergy?" *Gut* 57, no. 5 (2008): 561–67.

26. Nils Oskar Jõgi et al., "Zoonotic Helminth Exposure and Risk of Allergic Diseases: A Study of Two Generations in Norway," *Clinical & Experimental Allergy* 48, no. 1 (2018): 66–77. The idea of exposure to microbes as being protective has also been extended to parasites. There is a broad literature—including both scientific and popular— on the theory that at least part of our immune system evolved to counter the constant presence of a whole host of parasitic organisms in our natural environment. This theory—intimately related to the hygiene hypothesis—suggests that without parasites, the human immune system is prone to overreaction to other, less harmful substances. Yet new research directly counters the assumption that infection with intestinal parasites might be protective. Researchers at the University of Bergen in Norway found that children infected with helminths, common intestinal parasites, had quadruple the risk of developing asthma and allergies.

27. "Half of Ugandans Suffer from Allergy—Study," *The Independent*, July 25, 2019, https://www.independent.co.ug/half-of-ugandans-suffer -from-allergy-study/.

28. George Du Toit, M.B., B.Ch., Graham Roberts, D.M., Peter H. Sayre, M.D., Ph.D., Henry T. Bahnson, M.P.H., Suzana Radulovic, M.D., Alexandra F. Santos, M.D., Helen A. Brough, M.B., B.S., Deborah Phippard, Ph.D., Monica Basting, M.A., Mary Feeney, M.Sc., R.D., Victor Turcanu, M.D., Ph.D., Michelle L. Sever, M.S.P.H., Ph.D., et al., for the LEAP Study Team, "Randomized Trial of Peanut Consumption in Infants at Risk for Peanut Allergy," in *New England Journal of Medicine* 372 (2015): 803–813.

29. Victoria Soriano et al., "Has the Prevalence of Peanut Allergy Changed Following Earlier Introduction of Peanut? The EarlyNuts Study," *Journal of Allergy and Clinical Immunology* 147, no. 2 (2021). From a Melbourne study of 1,933 infants enrolled in the EarlyNuts study from 2018 to 2019 and compared to 5,276 infants in the HealthNuts study from 2007 to 2011. Guidelines were changed in 2016 to recommend

early introduction of peanuts and other allergenic foods before twelve months of age.

30. T. Feehley, C. H. Plunkett, R. Bao, et al., "Healthy infants harbor intestinal bacteria that protect against food allergy," in *Nature Medicine* 25 (2019): 448–453.

31. Brigham and Women's Hospital Press Release, "New Therapy Targets Gut Bacteria to Prevent and Reverse Food Allergies," June 24, 2019. https://www.brighamandwomens.org/about-bwh/newsroom/press-releases-detail?id=3352.

32. J. M. Anast, M. Dzieciol, D. L. Schultz, et al., "*Brevibacterium* from Austrian hard cheese harbor a putative histamine catabolism pathway and a plasmid for adaptation to the cheese environment," in *Scientific Reports* 9 (2019): 6164.

33. S. R. Levan, K. A. Stamnes, D. L. Lin, et al., "Elevated faecal 12,13-diHOME concentration in neonates at high risk for asthma is produced by gut bacteria and impedes immune tolerance," in *Nature Microbiology* 4 (2019): 1851–1861.

34. Emilie Plantamura et al., "MAVS Deficiency Induces Gut Dysbiotic Microbiota Conferring a Proallergic Phenotype," *Proceedings of the National Academy of Sciences* 115, no. 41 (2018): 10404–9.

35. Iweala and Nagler, "The Microbiome and Food Allergy."

36. Institut Pasteur, "Discovery of a Crucial Immune Reaction When Solid Food Is Introduced That Prevents Inflammatory Disorders," press release, March 19, 2019, https://www.pasteur.fr/en/press-area/press-documents/discovery-crucial-immune-reaction-when-solid-food-introduced-prevents-inflammatory-disorders.

37. One critique of the idea that antibiotics are the culprit is that this is correlation, and it may be that the infections are the real culprits and not the antibiotics themselves, especially since not all children who receive antibiotics will go on to develop allergies.

38. Zaira Aversa et al., "Association of Infant Antibiotic Exposure with Childhood Health Outcomes," *Mayo Clinic Proceedings* 96, no. 1 (2021): 66–77.

39. Joseph H. Skalski et al., "Expansion of Commensal Fungus *Wallemia mellicola* in the Gastrointestinal Mycobiota Enhances the Severity of Allergic Airway Disease in Mice," *PLOS Pathogens* 14, no. 9 (2018).

40. Anna Vlasits, "Antibiotics Given to Babies May Change Their Gut Microbiomes for Years," *STAT,* June 15, 2016, https://www.statnews.com/2016/06/15/antibiotics-c-sections-may-change-childs-health-for-the-long-term/.

41. Galya Bigman, "Exclusive Breastfeeding for the First 3 Months of Life

May Reduce the Risk of Respiratory Allergies and Some Asthma in Children at the Age of 6 Years," *Acta Paediatrica* 109, no. 8 (2020): 1627–33.

42. R. Bao et al., "Fecal Microbiome and Metabolome Differ in Healthy and Food-Allergic Twins," *Journal of Clinical Investigation* 131, no. 2 (January 19, 2021). Fecal study in infant twins suggests that differences in microbial populations in the gut—and also metabolites from dietary sources—might be responsible for food allergies. Changes in the gut microbiome persist into adulthood despite any changes in lifestyle factors or diet. The article quotes Kari Nadeau as saying that many people will go to Google wanting to know if eating yogurt is beneficial. And while they can't link these things to causation, there is a strong association. So, for now, there is no solid advice on what to eat.

43. Cheng S. Wang et al., "Is the Consumption of Fast Foods Associated with Asthma or Other Allergic Diseases?" *Respirology* 23, no. 10 (2018): 901–13.

44. Shashank Gupta et al., "Environmental Shaping of the Bacterial and Fungal Community in Infant Bed Dust and Correlations with the Airway Microbiota," *Microbiome* 8, no. 1 (2020): 115.

45. Technically, alpha-gal is a new food allergy—which we'll see later in this chapter—but it doesn't trigger the same allergic pathways. Thus, it both is and isn't a food allergy as we've come to think about them.

46. Samuel Feinberg, *One Man's Food* (Chicago: Blue Cross Commission, 1953): 6.

47. As an important aside, Paller doesn't agree with the classification of eczema as an allergic condition; in her opinion, it's lumped in with other allergies because of the allergic triggers. She also doesn't believe that the numbers show real evidence for the atopic march.

48. Iweala and Nagler, "The Microbiome and Food Allergy," 378.

49. Jaclyn Parks et al., "Association of Use of Cleaning Products with Respiratory Health in a Canadian Birth Cohort," *Canadian Medical Association Journal* 192, no. 7 (2020).

50. European Lung Foundation, "Exposure to Cadmium in the Womb Linked to Childhood Asthma and Allergies," ScienceDaily, September 2, 2020, www.sciencedaily.com/releases/2020/09/200902182433.htm. The children were followed up at age eight to see if they had any allergies.

51. Susanne Jahreis et al., "Maternal Phthalate Exposure Promotes Allergic Airway Inflammation over 2 Generations Through Epigenetic Modifications," *Journal of Allergy and Clinical Immunology* 141, no, 2 (2018): 741–53.

52. Allergies are also common in piglets weaned too early but remain extremely uncommon in cows.

53. Christine H. Chung, Beloo Mirakhur, Emily Chan, Quynh-Thu Le, Jordan Berlin, Michael Morse, Barbara A. Murphy, Shama M. Satinover, Jacob Hosen, David Mauro, Robbert J. Slebos, Qinwei Zhou, Diane Gold, Tina Hatley, Daniel J. Hicklin, Thomas A. E. Platts-Mills. "Cetuximab-induced anaphylaxis and IgE specific for galactose-α-1, 3-galactose," in *New England Journal of Medicine* 358, no. 11 (March 2008): 1109–1117.

54. This is why double-blind oral food challenges remain the gold standard of food allergy diagnosis. Neither the clinician nor the patient (or parents) may know if the patient has ingested the allergen or not. If either party knows, the results may be skewed. A top food allergist once told me that he's had patients absolutely swear they were allergic to something that they did not react to under the double-blind control challenge. It is not uncommon for food allergists to complain that their patients often do not want to accept the results of food challenges. The nocebo effect is strong enough for them to prefer their own evidence over that of the control trial.

55. Scott H. Sicherer and Hugh A. Sampson, "Food Allergy: Epidemiology, Pathogenesis, Diagnosis, and Treatment," *Journal of Allergy and Clinical Immunology* 133, no. 2 (February 2014).

56. U.S. Department of Health and Human Services, "Alpha-Gal Syndrome Subcommittee Report to the Tick-Borne Disease Working Group," last accessed February 13, 2022, https://www.hhs.gov/ash/advisory-committees /tickbornedisease/reports/alpha-gal-subcomm-2020/index.html.

Chapter 7: Remedies for the Irritated:
Allergy Treatments Past, Present, and Future

1. Samuel M. Feinberg, *Allergy in General Practice* (Philadelphia: Lea & Febiger, 1934).

2. Warren T. Vaughan, *Allergy and Applied Immunology: A Handbook for Physician and Patient, on Asthma, Hay Fever, Urticaria, Eczema, Migraine and Kindred Manifestations of Allergy* (St. Louis: C. V. Mosby, 1931). Vaughan details this discovery by Leonard Noon and John Freeman.

3. This practice echoes some modern-day alternative treatments that experiment with using parasites to tamp down the inflammation associated with negative immune responses. See Moises Velasquez-Manoff,

An Epidemic of Absence: A New Way of Understanding Allergies and Autoimmune Diseases (New York: Scribner, 2012).

4. Arthur F. Coca, *Asthma and Hay Fever in Theory and Practice. Part I: Hypersensitiveness, Anaphylaxis, Allergy* (Springfield, Ill.: C. C. Thomas, 1931), 744.

5. Arthur F. Coca, *Asthma and Hay Fever in Theory and Practice. Part I: Hypersensitiveness, Anaphylaxis, Allergy* (Springfield, Ill.: C. C. Thomas, 1931), 307–308.

6. George W. Bray, *Recent Advances in Allergy (Asthma, Hay-Fever, Eczema, Migraine, Etc.)* (Philadelphia: P. Blakiston's, 1931).

7. Fascinating aside: Tricyclic antidepressants also have antihistamine properties and are sometimes prescribed for hives. Antihistamines have also been found to help with nausea, vertigo, anxiety, and insomnia. Our bodies are more complex, and their systems are more interconnected, than I have space to dive into here.

8. Rachel G. Robison and Jacqueline A. Pongracic, "B Agonists," in *Patterson's Allergic Diseases,* 8th ed., ed. Leslie C. Grammer and Paul A. Greenberger (Philadelphia: Wolters Kluwer, 2018), 738.

9. Guy Laroche, Charles Richet, fils, and François Saint-Girons, *Alimentary Anaphylaxis (Gastro-intestinal Food Allergy)* (Berkeley: University of California Press, 1930), 125.

10. Albert Rowe, *Food Allergy: Its Manifestations, Diagnosis and Treatment, with a General Discussion of Bronchial Asthma* (Philadelphia: Lea & Febiger, 1931), 300–301.

11. Arthur F. Coca, *Asthma and Hay Fever in Theory and Practice. Part I: Hypersensitiveness, Anaphylaxis, Allergy* (Springfield, Ill.: C. C. Thomas, 1931), 270–310.

12. Christopher M. Warren et al., "Epinephrine Auto-injector Carriage and Use Practices Among US Children, Adolescents, and Adults," *Annals of Allergy, Asthma & Immunology* 121, no. 4 (October 2018): 479–89.

13. Dr. Jeanine Peters-Kennedy, an associate clinical professor at Cornell University's College of Veterinary Medicine, told me that pets typically receive allergen-specific immunotherapy (ASIT), though usually she just tells pet owners that these are vaccines for allergies. Once specific allergens are discovered and the immunotherapy is formulated, pet owners are trained to administer the shots at home. Unlike humans, pets don't have to come into a clinic to receive this type of therapy. Pets continuously take the shots. "They generally have it for life if it works. And it works in approximately two-thirds of the cases." They also get antihistamines, steroids on occasion, and other drugs to help with symptoms—just like us.

Allergy shots do seem effective for oral allergy syndrome, with 55 percent of children showing improvement in symptoms after receiving them. See "Allergy Shots May Be an Effective Treatment for Pediatric Pollen Food Allergy Syndrome," American College of Allergy, Asthma & Immunology, November 8, 2019, https://acaai.org/news /allergy-shots-may-be-effective-treatment-pediatric-pollen-food-allergy -syndrome.

14. Technical University of Munich (TUM), "Allergy Research: Test Predicts Outcome of Hay Fever Therapies," ScienceDaily, October 18, 2018, www.sciencedaily.com/releases/2018/10/181018095355.htm. A recent finding that in successful cases of immunotherapy treatment patients had more regulatory B cells and fewer TH-17 cells (a class of proinflammatory T helper cells) may lead to the development of a blood test that could predict the outcome of immunotherapy in patients. This would save patients who may not get as much benefit a lot of time and money.

15. New treatments on the horizon: LEO Pharma's new biologic tralokinumab, which blocks the IL-13 allergic pathway and was approved by the FDA in December 2021; Pfizer's PF-04965842, a daily oral Janus kinase (JAK)-1 enzyme blocker (the FDA labeled it as a "breakthrough therapy"); Eli Lilly and Incyte's baricitinib inhibits both JAK1 and JAK2.

16. Vittorio Fortino et al., "Machine-Learning-Driven Biomarker Discovery for the Discrimination Between Allergic and Irritant Contact Dermatitis," *Proceedings of the National Academy of Sciences* 117, no. 52 (2020): 33474–85.

17. "doc.ai Partners with Anthem to Introduce Groundbreaking, End-to-End Data Trial Powered by Artificial Intelligence on the Blockchain," PR Newswire, August 1, 2018, https://www.prnewswire.com/news -releases/docai-partners-with-anthem-to-introduce-groundbreaking -end-to-end-data-trial-powered-by-artificial-intelligence-on-the-block chain-300689910.html.

18. Kim Harel, "Researchers Describe Antibody That Can Stop Allergic Reactions," Aarhus University, January 28, 2018, https://mbg.au.dk /en/news-and-events/news-item/artikel/researchers-describe-antibody -that-can-stop-allergic-reactions/.

19. Donald T. Gracias et al., "Combination Blockade of OX40L and CD30L Inhibits Allergen-Driven Memory Th2 Reactivity and Lung Inflammation," *Journal of Allergy and Clinical Immunology* 147, no. 6 (2021): 2316–29.

20. Melanie C. Dispenza et al., "Bruton's Tyrosine Kinase Inhibition Ef-

fectively Protects Against Human IgE-Mediated Anaphylaxis," *Journal of Clinical Investigation* 130, no. 9 (2020): 4759–70.

21. BTK inhibitors are currently used in cancer treatments and cost about five hundred dollars per day. The possible downside? The drugs are known to cause immune system defects, leading to lower white blood cell counts and increased infections.

22. Julia Eckl-Dorna et al., "Two Years of Treatment with the Recombinant Grass Pollen Allergy Vaccine BM32 Induces a Continuously Increasing Allergen-Specific IgG4 Response," *The Lancet* 50 (November 27, 2019): 421–32.

23. Robert Heddle et al., "Randomized Controlled Trial Demonstrating the Benefits of Delta Inulin Adjuvanted Immunotherapy in Patients with Bee Venom Allergy," *Journal of Allergy and Clinical Immunology* 144, no. 2 (2019): 504–13.

24. American College of Allergy, Asthma, and Immunology, "Severe Eczema May Best Be Treated by Allergy Shots: Significant Benefits Seen in One Medically Challenging Case," ScienceDaily, November 16, 2018, www.sciencedaily.com/releases/2018/11/181116083213.htm.

25. "Animal Study Shows How to Retrain the Immune System to Ease Food Allergies," DukeHealth, February 21, 2018, https://corporate .dukehealth.org/news/animal-study-shows-how-retrain-immune -system-ease-food-allergies.

26. Northwestern University, "New Treatment May Reverse Celiac Disease: New Technology May Be Applicable to Other Autoimmune Diseases and Allergies," ScienceDaily, October 22, 2019, www.sciencedaily .com/releases/2019/10/191022080723.htm.

27. Jane AL-Kouba et al., "Allergen-Encoding Bone Marrow Transfer Inactivates Allergic T Cell Responses, Alleviating Airway Inflammation," *JCI Insight* 2, no. 11 (2017).

28. American Society of Agronomy, "Tackling Food Allergies at the Source," November 16, 2020, https://www.agronomy.org/news/science -news/tackling-food-allergies-source/.

29. The Chinese herbal remedy called "ma huang"—used for thousands of years for breathing troubles—led to the development of the drug ephedrine. In fact, many traditional herbs have been found to contain active compounds that have then been incorporated into Western biomedicine. Not all alternative or complementary treatments are bogus, though many are and can be dangerous if used without medical supervision. Some herbal preparations, for instance, have been found to contain dangerous substances like traces of lead.

30. Scott H. Sicherer and Hugh A. Sampson, "Food Allergy: Epidemiology,

Pathogenesis, Diagnosis, and Treatment," *Journal of Allergy and Clinical Immunology* 133, no. 2 (February 2014): 301. The authors state that "a 2012 World Allergy Organization review concluded that probiotics do not have an established role in the prevention or treatment of allergy."

When I interviewed Dr. Cathy Nagler, she told me, "The healthy microbiota is full of lactobacilli and bifidobacteria. These are the whole foods bacteria that you can get, typical probiotics. They don't work. They probably make your stomach feel better when it's upset, but they don't influence your immune system, except for some data for atopic dermatitis in the first year of life."

Chapter 8: The Booming Business of Allergy Treatments

1. In September 2016, the New York attorney general began an antitrust investigation into Mylan's practices as part of the EpiPen4Schools program. As a result, Mylan discontinued many of its sales practices related to the program.
2. Interesting sidenote: The patented injector mechanism that EpiPen uses was first invented in the 1970s by Sheldon Kaplan to deliver nerve gas antidote into soldiers. The EpiPen was first approved by the FDA in 1987.
3. Another pharmaceutical company, Novartis, entered the U.S. market with its own version to address EpiPen shortages in the summer of 2019.
4. For years, Mylan had avoided giving costly rebates owed to the government by falsely claiming that the auto-injectors it provided to patients under the program were generic versions and not a branded drug because they contained a generic and readily available drug—epinephrine.
5. The complaint argued that Mylan had been offering rebates to insurers and Medicaid if they agreed not to reimburse for Sanofi's auto-injector.
6. Interestingly, Mylan never formally admitted any wrongdoing in its pricing policies. The only person to ever take any public responsibility was Mylan CEO Heather Bresch at the 2016 Forbes Healthcare Summit. But even then, she claimed all the price hikes were justified by improvements the company had made to its products.
7. House bill 3435.
8. No financial sponsorship was provided by Sanofi and Regeneron to develop, publish, or distribute this book. The views and statements not directly quoting Regeneron or Sanofi employees are my own and not endorsed by Sanofi or Regeneron.
9. To be fair, there are side effects to every atopic dermatitis treatment. Extended use of corticosteroids can cause thinning skin, skin tears,

bruising, acne, rosacea, poor healing of wounds, and excess hair. However, most patients do well on steroids and most physicians attribute fear or paranoia about topical steroids to patient dissatisfaction with the treatments. I would add to this that if, as many people report, the steroids don't really help their condition that much, the minor side effects might not be worth all the trouble.

10. Sarah Faiz et al., "Effectiveness and Safety of Dupilumab for the Treatment of Atopic Dermatitis in a Real-Life French Multicenter Adult Cohort," *Journal of the American Academy of Dermatology* 81, no. 1 (July 1, 2019): 143–51.

11. G. A. Zhu et al., "Assessment of the Development of New Regional Dermatoses in Patients Treated for Atopic Dermatitis with Dupilumab," *JAMA Dermatology* 155, no. 7 (2019): 850–52.

12. Trying to keep up with current drug prices throughout the researching, writing, and editing of this book pushed me to the brink of sanity. Prices fluctuate frequently and actual cost to the patients varies wildly depending on a host of factors. If you want to see the current price as you're reading this, do a web search. It will likely be in the ballpark of this number but predicting what drug prices will be at any given moment is like predicting the weather during climate change.

13. Regeneron's CEO, Len Schleifer, has openly said that he sees a heavy marketing push as the key to increasing the revenue for Dupixent, turning a $2 billion per year drug into a $12 billion per year drug.

14. "Air Purifier Market Share, Size, Trends, Industry Analysis Report by Type [High Efficiency Particulate Air (HEPA), Activated Carbon, Ionic Filters]; by Application [Commercial, Residential, Industrial]; by Residential End-Use; by Region, Segment Forecast, 2021–2029," Polaris Market Research, November 2021, https://www.polarismarket research.com/industry-analysis/air-purifier-market.

15. For more on this trend, see https://www.pureroom.com/; Tanya Mohn, "Sneeze-Free Zone," *New York Times,* January 10, 2011, https://www .nytimes.com/2011/01/11/business/11allergy.html; and Alisa Fleming, "Hotel Havens for Travel with Allergies and Asthma," *Living Allergic,* February 5, 2014, https://www.allergicliving.com/2014/02/05/hotel -havens/.

16. To counter this, the Asthma and Allergy Foundation of America started a program to vet such products. Their method and list of products can be found at https://www.asthmaandallergyfriendly.com/USA/.

Chapter 9: What Makes a Treatment Effective?
Weighing Benefits and Risks

1. U.S. Food and Drug Administration, "Benefit-Risk Assessment in Drug Regulatory Decision-Making," March 30, 2018, 3, www.fda.gov/files /about%20fda/published/Benefit-Risk-Assessment-in-Drug-Regulatory -Decision-Making.pdf.

2. The 2014 edition of the American Academy of Allergy, Asthma & Immunology's *Practice Management Resource Guide* features an entire chapter on how to mix standardized allergens to individualized immunotherapy kits for patients.

3. Thomas Casale, A. Wesley Burks, James Baker, et. al, "Safety of Peanut *(Arachis hypogaea)* Allergen Powder-dnfp in Children and Teenagers With Peanut Allergy: Pooled Analysis from Controlled and Open-Label Phase 3 Trials," in *Journal of Allergy and Clinical Immunology* 147, no. 2 (2021): AB106.

4. Palforzia treatment currently doesn't exceed 300 milligrams. If patients want to continue with OIT, as Stacey did, they will have to stop Palforzia and switch to more traditional OIT.

5. R. Chinthrajah et al., "Sustained Outcomes in a Large Double-Blind, Placebo-Controlled, Randomized Phase 2 Study of Peanut Immunotherapy," *Lancet* 394 (2019): 1437–49.

6. Institute for Clinical and Economic Review (ICER), "Oral Immunotherapy and Viaskin® Peanut for Peanut Allergy: Effectiveness and Value Final Evidence Report," (July 10, 2019): ES6.

7. Institute for Clinical and Economic Review (ICER), "Oral Immunotherapy and Viaskin® Peanut for Peanut Allergy: Effectiveness and Value Final Evidence Report," (July 10, 2019): ES6.

8. Institute for Clinical and Economic Review (ICER), "Oral Immunotherapy and Viaskin® Peanut for Peanut Allergy: Effectiveness and Value Final Evidence Report," (July 10, 2019): ES7.

9. This is one of the strengths—and weaknesses—of advocacy groups for specific medical conditions. While basic science is critical to the development of treatments, advocacy groups often push for more targeted, applied research instead. This can be good, in the sense that it does streamline the process, but it is also bad, in the sense that it takes funding away from our understanding of the mechanisms that underlie all allergic responses. It's difficult to balance the need for more basic science with patients' desires for more applied science—in the form of new treatments. Advocacy groups often traffic in "wishful thinking" as well and sometimes push the approval of drugs that aren't as effective

as we would hope. For a pertinent example of this, please see coverage of the controversial new drug for Alzheimer's: Pam Belluck, "Inside a Campaign to Get Medicare Coverage for a New Alzheimer's Drug," *New York Times,* April 6, 2022, https://www.nytimes.com/2022/04/06 /health/aduhelm-alzheimers-medicare-patients.html.

10. Press Release, "Nestlé to Acquire Aimmune Therapeutics," Aug. 31, 2020, https://www.nestle.com/media/pressreleases/allpressreleases/nestle -to-acquire-aimmune-therapeutics.

11. The list of criticisms against Nestlé is legion and covers everything from the use of child and slave labor to pollution. For a good rundown, see https://www.zmescience.com/science/nestle-company-pollution-children/ and https://www.mashed.com/128191/the-shady-side-of-mms/ and https://www.ethicalconsumer.org/company-profile/nestle-sa.

12. In fact, in 2019, Nestlé invested in another food allergy therapeutics company called SpoonfulONE. One of its founders, Dr. Kari Nadeau, is a top food allergist. The company was founded entirely by women and aims to prevent food allergies from developing via early delivery of sixteen major food allergens.

13. K. Papp et al., "Efficacy and Safety of Ruxolitinib Cream for the Treatment of Atopic Dermatitis: Results from 2 Phase 3, Randomized, Double-Blind Studies," *Journal of the American Academy of Dermatology* 85, no. 4 (October 2021): 863–72.

14. Institute for Clinical and Economic Review (ICER), "JAK Inhibitors and Monoclonal Antibodies for the Treatment of Atopic Dermatitis: Effectiveness and Value Final Evidence Report," August 17, 2021, https://icer.org/wp-content/uploads/2020/12/Atopic-Dermatitis_Final -Evidence-Report_081721.pdf.

Chapter 10: Allergy Is a Social Problem, Too

1. L. Gibson-Young, M. P. Martinasek, M. Clutter, and J. Forrest, "Are Students with Asthma at Increased Risk for Being a Victim of Bullying in School or Cyberspace? Findings from the 2011 Florida Youth Risk Behavior Survey," *Journal of School Health* 87, no. 7 (July 2014): 429–34.

2. Eyal Shemesh et al., "Child and Parental Reports of Bullying in a Consecutive Sample of Children with Food Allergy," *Pediatrics* 131, no. 1 (2013).

3. American College of Allergy, Asthma, and Immunology, "Nearly One in Five Parents of Food-Allergic Children Are Bullied," ScienceDaily (November 13, 2020), www.sciencedaily.com/releases/2020/11/201113 075250.htm.

4. The survey was performed by NORC at the University of Chicago (formerly the National Opinion Research Center), an independent social science research agency. NORC is well known for its rigorous methodology. As such, the survey participants accurately reflect the current demographic makeup of the American population per the last U.S. census. While survey results are, at best, a snapshot of public opinion, I feel confident arguing that NORC provided me with the best possible data on American attitudes and beliefs about allergy.

5. Interesting sidenotes: By far the largest constituent of people who responded yes to having an allergy reported that they suffered from hay fever. One in four survey respondents said they have it. In addition, 39 percent of those who responded yes were self-diagnosed. They have never seen an allergist or other healthcare professional about their allergies.

6. Those who did tended to have a high school diploma or less education.

7. In the middle of our empathy scale are asthma and type 2 diabetes. They ended up ranked in the middle of most people's lists.

8. Elizabeth DiFilippo, "Mother's heartbreaking warning after daughter with peanut allergy dies from eating cookie," in *Yahoo! Finance News,* July 18, 2018. https://finance.yahoo.com/news/mothers-heartbreaking -warning-daughter-peanut-allergy-dies-eating-cookie-2-140139277 .html.

9. P. Joshi, S. Mofidi, and S. H. Sicherer, "Interpretation of Commercial Food Ingredient Labels by Parents of Food-Allergic Children," *Journal of Allergy and Clinical Immunology* 109, no. 6 (June 2002): 1019–21.

10. Sarah Besnoff, "May Contain: Allergen Labeling Regulations," *University of Pennsylvania Law Review* 162, no. 6 (May 2014): 1465–93.

11. M. J. Marchisotto et al., "Food Allergen Labeling and Purchasing Habits in the United States and Canada," *Journal of Allergy and Clinical Immunology: In Practice* 5, no. 2 (March–April 2017): 345–51.

Epilogue: Irritating Ourselves to Death: Allergy in the Time of COVID-19

1. René Dubos, *Man and His Environment: Biomedical Knowledge and Social Action* (Washington, D.C.: Pan American Health Organization/ World Health Organization, 1966):168.

2. University of Exeter. "The 'pathobiome'—a new understanding of disease." ScienceDaily. www.sciencedaily.com/releases/2019/09/190912 113238.htm (accessed August 26, 2022).

SUGGESTIONS FOR FURTHER READING

Braun, Lundy. *Breathing Race into the Machine: The Surprising Career of the Spirometer from Plantation to Genetics*. Minneapolis: University of Minnesota Press, 2014.

Jackson, Mark. *Allergy: The History of a Modern Malady*. London: Reaktion Books, 2006.

Mitman, Gregg. *Breathing Space: How Allergies Shape Our Lives and Landscapes*. New Haven, Conn.: Yale University Press, 2007.

Sicherer, Scott H. *Food Allergies: A Complete Guide for Eating When Your Life Depends on It*. Baltimore: Johns Hopkins University Press, 2013.

Smith, Matthew. *Another Person's Poison: A History of Food Allergy*. New York: Columbia University Press, 2015.

Suggestions for More Information and Getting Involved

Stacey Sturner's Facebook group:
https://www.facebook.com/groups/foodallergytreatmenttalk/

Emily Brown's Food Equality Initiative:
https://foodequalityinitiative.org/

National Eczema Association:
https://nationaleczema.org/

Food Allergy Research & Education (FARE):
https://www.foodallergy.org/

Asthma and Allergy Foundation of America:
https://www.aafa.org/

ABOUT THE AUTHOR

Dr. Theresa MacPhail is a medical anthropologist (PhD University of California, Berkeley, and University of California, San Francisco), a former journalist, and an associate professor of Science and Technology Studies at Stevens Institute of Technology. She researches and writes on topics in global health, biomedicine, and disease. Dr. MacPhail grew up split between a double-wide trailer in rural Indiana and a nice home in rural New Hampshire. She remains a proud Hoosier who tries her best to live free and avoid dying. She lives in Brooklyn with her old cat and billions of dust mites.

ABOUT THE TYPE

―――――

This book was set in Sabon, a typeface designed by the well-known German typographer Jan Tschichold (1902–74). Sabon's design is based upon the original letterforms of sixteenth-century French type designer Claude Garamond and was created specifically to be used for three sources: foundry type for hand composition, Linotype, and Monotype. Tschichold named his typeface for the famous Frankfurt typefounder Jacques Sabon (c. 1520–80).